ELECTRICITY
FOR
RURAL
AMERICA

ELECTRICITY
FOR
RURAL
AMERICA

The Fight for
the REA

D. CLAYTON
BROWN

Contributions in Economics and Economic History, Number 29

Greenwood Press
Westport, Connecticut . London, England

Library of Congress Cataloging in Publication Data

Brown, Deward Clayton, 1941-
    Electricity for rural America.

    (Contributions in economics and economic history ; no.
29 ISSN 0084-9235)
    Bibliography: p.
    Includes index.
    1. Rural electrification--United States--History.
2. United States. Rural Electrification Administration--
History. I. Title.
HD9688.U52B76   353.008'72208'09   79-8287
ISBN 0-313-21478-6

Library of Congress Catalog Card Number: 79-8287
ISBN: 0-313-21478-6
ISSN: 0084-9235

First published in 1980

Greenwood Press
A division of Congressional Information Service, Inc.
88 Post Road West, Westport, Connecticut 06880

Printed in the United States of America

10 9 8 7 6 5 4 3 2

CONTENTS

TABLES

PREFACE

Extending electricity to farms and rural homes was an uphill struggle that lasted for two generations. Foremost among the obstacles to be overcome was the refusal of the electrical industry to serve the rural market and its opposition to the development of public power as an alternate means of serving the farmer. Rural electrification, therefore, was part of a larger question over private versus public development of energy resources, and it had to be resolved in the political arena.

The creation of the Rural Electrification Administration (REA) in 1935 meant that the electrification of agriculture was accepted as a public responsibility. But it was still no easy task to serve every farm and home since the new agency encountered stiff resistance from utility companies, and had to devote considerable attention to organizing its own administration. Closely related to the organizational task was the establishment of a policy over matters such as power generation and subsidization. Not until the end of World War II were these fundamental questions resolved. In the development of rural electrification, therefore, struggles within the internal structure of REA played an important role.

At the end of the war, shortages of electricity prevented the full development of rural service, mostly in the South. A postwar search for new sources of energy in the region was made independently of REA, although the purpose was to supply cooperatives with ample energy. Only when REA was fully established as a lending and supervisory agency and new sources of energy were developed could the United States claim to have complete electrification.

Once that goal was reached, rural life improved remarkably. The benefits of electrical technology such as running water, refrigeration, radio, and sanitation lifted farm families out of the preindustrial life and enabled them to enjoy a lifestyle conforming to the standards of an industrialized society. The internal combustion engine contributed to this transformation, but electricity provided not only mechanized power for farming operations but also improved school and community life. Best of all, electricity could be brought into the home for personal use. Because of its versatility, electric service was the single most important development responsible for ending the drudgery and toil of farm life and alleviating the generally depressed physical and spiritual condition of farmers.

Although the push for public rural electrification was nationwide, the far western states made the least contribution. They had natural conditions that made farm service profitable for power companies: use of irrigation, large-scale farming, specialized production and a low proportion of family "dirt farms." By 1935 when REA went into operation, the three Pacific states—California, Oregon and Washington—averaged more than 50 percent rural electrification; California alone had 63 percent of its farms served. The Midwest and South, where conditions did not encourage power companies, had a low percentage, which explains their active role in public electrification. In 1930 the North Central states averaged 13 percent electrification, and the South Central had only 3 percent.

Because of this difference in regional influences, caution was used in selecting topics for discussion in this book. I have discussed only those cases in which electricity for agriculture was the central point of argument, for not all public power debates were concerned with rural electrification. To be included herewith a topic had to bear directly on the subject and also contribute to the broader sweep of events. Incidents that were only peripheral to the main thrust of the crusade to serve farms have been excluded. For guidance in this respect, reliance was placed on primary materials.

REA proved to be one of the most successful agencies started in the 1930s. Still in operation today, REA has 979 cooperatives and feeds comfort-giving current to 7,767,813 households. A remarkable feature has been the success of the self-liquidating concept, meaning that the program paid for itself with less than 1 percent default on loans since 1935. In 1949 Congress authorized the agency to provide telephone service for the millions receiving its service. This work has focused solely on electrification, however, for telephone service came anticlimactically and did not engender the controversy nor have the impact of electricity on rural life.

I wish to express my appreciation to those individuals who assisted me. Without their contribution this book would have been impossible. To my mentor, Theodore Saloutos of the History faculty at the University of California, Los Angeles, I owe a debt for his patience and perserverance. To G. C. Hoskins of Southern Methodist University, I owe an equally large debt for his generous support and editorial criticisms. To my colleague, Donald Worcester, I am indebted for his keen editorial pen. Roger Daniels of the University of Cincinnati donated his talents for organization and structure to the manuscript. Richard Lowitt of the University of Iowa provided information on George Norris and the subject of public power. Harold T. Pinckett furnished valuable information on the Gifford Pinchot papers.

I would like to thank Columbia University's Oral History Research Office for permission to quote from "The Reminiscences of Samuel B. Bledsoe," "The Reminiscences of John Carmody," and "The Reminiscences of Edward O'Neal" (each copyright © 1972 by the Trustees of Columbia University in the City of New York).

I was served splendidly by librarians in my quest for research materials. H. G. Dulaney, director of the Sam Rayburn Library, holds a special place in this respect. Substantial help also came from the personnel at the National Archives, Library of Congress; the National Agricultural Library; the Franklin D. Roosevelt Library; the Herbert Hoover Library; the TVA Library at Knoxville; the Columbia University Library; the Duke University Library and the Texas Christian University Library. A special note of thanks must go to the Alabama Power Company which granted me full access to their library. The officers and staff of the National Rural Electric Cooperative Association furnished hard-to-find data, and the REA Information Service provided assistance and information for which I am also grateful. I am also indebted to those scholars whose work on rural electrification preceded mine. Without their contributions, my own study probably would not have been possible.

Several individuals whose lives were part of the story of rural electrification gave their valuable time for interviews. Sharing their memories were Clyde T. Ellis, long-time executive manager of the National Rural Electric Cooperative Association; Congressman Robert Poage of Texas; Douglas G.

Wright, first administrator of the Southwestern Power Administration; E. C. Easter of the Alabama Power Company; and Truett Bailey of the Brazos River Generating and Transmission Cooperative. The contributions of each added to my understanding of the numerous incidents and events that enabled rural inhabitants to enjoy the benefits of electrical technology.

Funding was provided by the Texas Christian University Research Foundation and the National Endowment for the Humanities. Their generous assistance speeded up the date of completion.

My wife and children made more than their share of sacrifices on behalf of this work. Their love, devotion and patience were plentiful as well as essential.

D. Clayton Brown
Texas Christian University
October, 1979

INTRODUCTION

Prior to World War II, rural homes generally had few of the technological comforts and conveniences synonomous with modern living. Families experienced a life filled with toil and drudgery, relying on hand labor and animal power. Sanitation in the home was poor, and disease and ill health due to that condition were chronic in some areas. Automobiles and tractors had brought some of the advantages of technology to the farm, but the single most important ingredient required for modern living, electricity, was not available. Only one out of ten farms in the United States had electric service.

Keeping house, washing clothes and tending to chores without modern appliances were monotonous and accounted for much of the hardship. Farmers typically had no running water and indoor bathrooms. They had to bring water from streams or pump it from a nearby well. In 1919 the United States Department of Agriculture (USDA) reported that rural families spent over 10 hours per week pumping water and carrying it from source to kitchen.[1] A time-consuming chore, this practice discouraged cleanliness and caused country doctors to carry distilled water and their own towels when visiting patients.

Washing the family laundry was a burdensome task since clothes usually were scrubbed in an outdoor tub or kettle filled with water heated by a fire. Some women had hand-operated washing machines, but the kettle and washboard prevailed. Use of the "sad iron," an iron heated on a wood-burning stove, made the laundry still more debilitating. According to one report, farmwives spent twenty days more per year washing clothes than women in the city using electric washers.[2]

The lowly privy, however, was the principal source of hardship in view of its contribution to ill health. Frequently the privy contaminated the water supply, causing typhoid, dysentery and a variety of gastrointestinal illnesses. The Chief of the Maryland Department of Public Health reported that dysentery was the principal cause of a high infant mortality rate in the state's rural environs. Hookworm was a by-product of the outdoor toilet in the South where weather and moist soil created ideal conditions for the larvae. Rate of hookworm infestation in southern schools went to 50 percent, or higher. This parasite caused loss of energy, weakness and general anemia, and for good reason it was called the "germ of laziness." It was well-established that the privy was the causal agent of hookworm; in those areas "where hookworm is prevalent," wrote one health officer, "toilet facilities are of the most primitive kind."[3]

Poor health cannot be attributed solely to the absence of running water and use of the privy. Suffering was also related to the improper storage of food. Until farmers had electricity, they preserved food much like their ancestors, relying on block houses, smokehouses, cellars or a combination thereof. Perishable commodities such as milk, butter, eggs and fresh fruit were put in a well or springhouse to cool, and this method of refrigeration, with its obvious limitations, caused much waste, especially during hot weather. Spoilage of food, an everyday occurrence on the farm, was also responsible for dysentery, undulant fever and gastrointestinal disturbances.

In the South where the prolonged warm weather made preservation of food even more difficult, the tenant or sharecropper class had a notoriously deficient diet of salted fatback, cornmeal and molasses, which required no refrigeration. Dangerously lacking in essential vitamins, iron and protein, prolonged consumption of these foods caused chronic fatigue and general debility, symptoms which accounted in part for the reputation of tenants as a shiftless and lazy people. This diet was also responsible for pellagra, which was so common in the South that the region was known as the "pellagra belt." In advanced stages, pellagra produces skin lesions, bleeding and mental aberrations, and although not as deadly as some diseases, it killed some 3,000 southerners in 1935 alone. A large

number of southerners were "prepellagrins," having only the first symptoms of scaly and inelastic skin, and a quality of carelessness and loss
of energy.[4]

Unbalanced diets struck in still more ways. By World War II it was already established that poor nutrition during pregnancy could cause still-
born infants, premature or difficult delivery, congenital deformities and
a host of pregnancy-related toxemias. With increasing evidence that a
child's intelligence and behavioral growth are directly related to nutrition,
rural electrification and refrigeration took on greater importance every
day. Quality of diet corresponded to income, however, and the worst
features of the poor diet usually were found among blacks, and white
sharecroppers in the South. Therefore, lack of refrigeration cannot be
considered the sole cause of diet-related illness, but the two are connected.[5]

Lack of electrical equipment outside the home also forced the farmer
to conduct his operations under strenuous conditions. Although by 1930
gasoline tractors had begun to reduce the labor and time of field work
for some, most farmers still relied on muscle power for mundane tasks
such as drawing water, feeding livestock, making repairs and a host of
chores that electrical machinery could perform more efficiently. Continued use of the kerosene lantern in the home and barn exemplified the
preindustrial conditions on a large segment of American farms.

The list of monotonous farm and home chores conducted by hand
was quite long. By 1930 agricultural engineers had developed 200 applications of electricity to farm life. To the extent that the lack of electric
service kept rural people from enjoying the benefits of industrialization,
they continued to follow nineteenth century guidelines for judging their
standard of living and regarded the expenditure of energy as the criterion
for "honest work."

As the use of modern conveniences became common in the cities, however, rural youth was unwilling to accept the old standards for judging
the quality of life, and they began migrating to the cities. When the federal
census of 1920 showed a growing preference for urban life, it also reported that of the total 6,000,000 farms in the United States, only
452,620 had electric lights and 643,899 had some form of running water.
The latter included cisterns and use of gravity-induced flow of water into
the house. Most of the farm homes with electricity were concentrated in
New England and the far West where the number of serviced farms ranged

from 15 to 45 percent respectively. The Midwest and South ranked lowest, ranging from 10 percent to less than 1 percent. By 1930 some improvement had occurred in the West, and California had more than 60 percent electrification, but no appreciable gains were made in the Midwest or South where the percentages remained approximately the same.[6]

Preference for urban life came as a shock to a country that had long regarded farmers as the backbone of the republic. Fear for the loss of the rural way of life was widespread, and agriculturalists saw a connection between abandonment of the land and lack of comforts in the home. "Not until farms and houses are equipped with the conveniences of modern life," wrote one observer, "will any considerable body of people be content to endure the hardships and loneliness of the rural sections."[7] Gifford Pinchot, a well-known political and conservation spokesman, went even further: "Only electric service can put the farmer on an equality with the townsmen and preserve the farm as the nursery of men and leaders of men that it has been in the past."[8] Therefore, the lack of electricity meant more than inconvenience; it was viewed as a chief cause of the decline of the rural way of life.

It was in the context of the preservation of rural culture as well as the need for farm modernization that electrification came to be regarded as essential. A monumental barrier of economic, technological and organizational problems, however, was involved in serving farms. It was in the early 1920s when the power companies were regarded as the appropriate instigators of service to the rural market that the first attempts were made to solve these problems.

1
THE FARMER
AND THE
ELECTRICAL INDUSTRY

Until the creation of the Rural Electrification Administration (REA) in 1935, power companies had the prerogative to serve farmers, but they were slow or unwilling to do so because of the high cost involved. Hoping to reduce cost and extend service on a broad scale, the electrical industry in 1923 began a special cooperative program with state agricultural colleges and the American Farm Bureau Federation (AFBF). The program was known as the Committee on the Relation of Electricity to Agriculture (CREA). For nearly a decade attention focused on CREA, but it was only a half-hearted attempt, too lacking in commitment to succeed. Not until CREA had run its full course, however, was it clear that the farmer could not rely on the power industry for service.

The National Electric Light Association (NELA), trade guild of the power industry, had first taken up the subject of rural electrification in 1911 at its annual meeting. Suggestions were made that more attention should be paid to agriculture. "The subject of electricity in rural districts," it was stated, "is one which the Central Station industry has only recently concerned itself with."[1] A special committee of NELA asked USDA to publish a bulletin on the farm uses of electricity, and it also requested that the Census Bureau secure data on the number of farms already served.[2]

No plans to promote service were implemented either by the NELA or the power companies acting alone. In 1913, the USDA published a special bulletin on farm uses of electricity, but otherwise took no action.

The power industry, however, should not be indicted for showing little interest prior to World War I. Electricity was considered a luxury in rural areas, a servant in the home or an extra hand at chores. Farmers placed emphasis on hard work and long hours and hardly thought of having modern appliances on their premises. But as industrialization spread and as electricity with its many benefits became a regular feature in city homes, attitudes changed rapidly, and by the 1920s the rural use of electricity was no longer thought of as a luxury, but as a technological advancement necessary to fulfill the promise of American life.

A more significant step to serve farms came in 1923 when NELA created CREA. Its origins went back to the 1921 meeting of NELA when the Overhead Systems Committee, discussing the construction of cross-country transmission lines, reported that company managers looked upon the rural market as too troublesome and seldom bothered with it. If the companies refused to serve farmers, the committee warned, "they will provide it for themselves." Other topics at the same meeting dealt with "the insistent demand for the conveniences which were enjoyed only by the residents of the city sections."[3] As the result of this urging by its brethren, NELA formed the Rural Lines Committee to look further into the matter.

The new committee made several recommendations to speed development. Since there was little knowledge of the potential agricultural uses of electricity, it called for a program of research and experimentation. Farmers also would have to be educated in the uses of electricity, then persuaded that it was profitable and beneficial to their operations. More important, however, was the committee's recognition of the high cost of service. Rates, they recommended, would have to be reduced to be within the reach of the rural homeowner. Only when these barriers were overcome could service be extended on a large scale.[4]

Requiring the assistance of agricultural interests, representatives of NELA met with officials of AFBF on September 11, 1922. After several sessions, the two groups agreed that they needed help from the various parties that might be affected by their plans, and they invited representatives of USDA to join the meetings. To promote coordination of the power industry and agriculture, they formed CREA in March 1923, with J. W. Coverdale of the Farm Bureau serving as president and Grover C. Neff of the Rural Lines Committee and executive officer of Wisconsin

Power and Light as secretary. The newly formed organization announced as its goal the electrification of one million farms by 1935, but its ultimate purpose was to remedy the lack of rural service.[5]

Even though membership included public interests, CREA was oriented toward private enterprise. Funds for the program came from individual power companies and appliance manufacturers. At no time during its operation did CREA seek the advice of public power proponents or consumer affairs groups. E. O. Bradfute, Farm Bureau president, stated that "farmers believe in private ownership of public utilities . . . rather than municipal, state, or federal ownership."[6] The USDA exercised no policy-making leadership and demonstrated no desire to represent anything other than the power industry point of view; USDA, if anything, was a surrogate partner and not an innovator. Industry's preeminence reflected the popular belief in 1923 that only private enterprise should develop the agricultural market.[7]

Cost was the real stumbling block to service. Rural lines cost $2,000 or more per mile, and since there were usually only two to five dwellings per mile in the country, utilities anticipated low revenue to amortize investments. They preferred the urban market. Companies expected farmers, therefore, to bear the burden of the initial investment charging them with the cost of the line, or a $500 to $1,000 deposit. Rural rates were also high, about 9 to 10 cents per kilowatt-hour for the minimum usage. No such adverse conditions applied to city dwellers who paid 4 to 5 cents per kilowatt-hour and were under no obligation to pay for the cost of the line.[8]

Few rural homeowners could afford to pay for the lines or make the deposit, nor could they at first afford enough appliances to use the amount of electricity necessary to achieve the advantage of lower rates. The effect was an endless cycle of expense for both parties—recipients of service used little power because of high rates, and the utilities charged such rates because of low usage. After two or three years, families usually had enough appliances to enable them to use more energy, but the waiting time was a capital risk the industry was unwilling to take. Greater use of energy, furthermore, was no assurance to farmers that they would receive lower rates.

The CREA plan put the responsibility for overcoming these barriers on the customer. Although research would be conducted on developing uses of electricity, and farmers shown how to apply them, consumers were expected to use a maximum of energy so the cost per kilowatt-hour would drop within their budget. Electric companies, sniffing new opportunities, would hopefully extend service. As the plan was designed, the solution was through increased sales; rates would be lowered only with increased use of power, which explains the emphasis on maximum use. Critics of the power industry scoffed at the assumption that the solution lay in research and consumer education, asserting that farmers already knew how to use electricity. They insisted that cost was the chief drawback.

Strategy of CREA was based on single state committees consisting of an electric company, the state chapter of the Farm Bureau and the state agricultural college. Experiments on the use of energy were conducted on campus farms and the results published in bulletins, farm and engineering journals and related communications. The company served a selected area, a "laboratory community of farms," so that the practical use of electricity could be observed. Officers of the Farm Bureau acted as an expediting force, assisting in recruiting participants, relaying information and publicizing the progress and results of the program.[9]

Altogether thirty-one state committees were created during the lifetime of CREA. A variety of experiments on the application of electricity to farming were conducted, from pumping water to heating seedbeds for crops, from milking cows to watering poultry. Valuable technical progress was made and farmers in the respective states were notified of the results through public displays and fairs, county agents, the mail and other avenues of rural communication.[10]

The CREA projects were conducted under conditions, however, that made them unrealistic, and the classic example was at Red Wing, Minnesota, a small dairying community of nineteen families living near a power line. The Northern States Power Company built a six-mile extension to the community, and a host of appliance manufacturers equipped the premises with their products. E. A. Stewart, professor of agricultural engineering at the University of Minnesota, served as director. Lights were put in the houses and barns, and a meter attached to every appliance, making it possible to determine the exact power usage. Families kept records of the cost and time saved. Use of appliances reduced the drudgery and time spent at backbreaking chores, and with greater consumption of energy the cost per kilowatt-hour was also reduced. Profits of the dairies were not increased, but the electricity paid for itself. After the study ended, Red Wing dairymen kept their service for the sake of convenience if for no other reason.[11]

NELA proudly pointed to Red Wing as an example of its efforts to promote electrification and publicized the project widely. Numerous bulletins and reports on the experiment were sent to agricultural colleges, the farm press and radio, power companies, engineering societies and other interested parties. Power companies looked upon the results of the study as proof that rural use of electricity was profitable and that farmers deserved the blame if they lacked service. Utility executives nationwide used Red Wing as a defense against any criticism leveled at them for the lack of service.[12]

Although successful in showing how power could be used, Red Wing had serious drawbacks for testing the feasibility of nationwide electrification. It was an experiment conducted in an artificial situation since the participants had special advantages conducive to the use of energy: small acreage, high use of machinery, close proximity to an existing power line and regular income from dairy production. Typical farms in the United States were more widely dispersed and less specialized. The average farmer used less machinery, and his income was dependent on the hazardous seasonal harvest. The rate structure of the Minnesota experiment, although based on a graduated scale, still left the cost too high for more typical farming operations, even if Red Wing participants could afford it. In 1927 the average rate at Red Wing came to 6.46 cents per kilowatt-hour, an improvement, but still too costly. "The results," wrote one critic, "failed to impress agriculture."[13]

A true test of the plan would require use of power by real "dirt farmers" conducted on a scale large enough to be more than a "laboratory community." The only CREA project established in such manner was in Alabama. It not only furnished an opportunity to determine if farmers could afford power at the rates charged at the time but also provided a chance to appraise the social ramifications of electrification on bona-fide "dirt farms."

The Alabama CREA grew out of discussions among Thomas W. Martin, president of the Alabama Power Company, representatives of the Department of Agricultural Engineering at Alabama Polytechnic Institute and

Edward O'Neal, president of the Alabama Farm Bureau Federation. Determined to have more than a laboratory study, they arranged to extend service to a large number of ordinary farms and homes, believing that farmers could be expected to require service only if electricity were profitable under regular field conditions. "To put up a line or two and study conditions," they predicted, "will never solve the rural distributor's problems."[14]

The project started when engineers at Auburn searched for ways to use electricity in farm work. Of special note was an attempt to design a low-cost, combination refrigerator-freezer. Need for refrigeration was self-evident given the hot weather of the region which made the storage of perishable foods almost impossible. They tried to design an economical unit big enough to freeze large quantities of food and at the same time refrigerate smaller quantities. Unfortunately this particular part of the study was unsuccessful, but it reflected an effort to cope with problems of the rural family.[15]

More important, however, were the observations made of families using power at home. The Alabama Power Company selected several counties as test sites and extended service to willing participants. No special arrangements were made to entice them to use more than the normal amount of service. Recipients agreed to keep records of their uses of power, but otherwise were free to follow their own inclinations. Participating were 379 farms, most of which produced cotton and grain. Dairy and truck farms and a few miscellaneous agricultural enterprises accounted for the rest. A large proportion of the participants were simply classified as "other rural customers," meaning families not wholly dependent on farming and small town residents. Included were eighty-five schools, churches and lodges. Altogether there were 1,880 units, and as M. L. Nichols, chairman of the Agricultural Engineering Department at Alabama Polytechnic Institute, commented, "this gives the work an entirely different aspect from the purely experimental work on a theoretical basis."[16]

During the course of the Alabama project, which lasted three years, valuable observations were made in the field work involved in recruiting, organizing and securing the cooperation of the families for the study. Participants were eager for service, but there remained the obvious tasks of helping them wire their homes, and teaching them how to use electrical equipment and to keep records of such use. Once families had agreed to take part, community meetings were held at which final decisions were reached. This face-to-face contact was particularly effective in giving participants confidence in the program.[17]

As the Alabama project progressed it was also clear that electricity had greatest impact in easing the burden of keeping house. Wives mentioned freedom from carrying water and caring for kerosene lamps as their most prized releases from drudgery. Participants devoted more time to evening activities such as reading and listening to the radio. Families with refrigerators welcomed the availability of a greater quantity of fresh foods as well as the convenience of cold storage. Electric service, besides reducing toil, increased the participants' self-satisfaction because it brought tangible evidence of modernization. The benefits would, according to O'Neal, "result in a higher standard of living which cannot be figured in dollars and cents or in kilowatts."[18]

Participants also utilized electricity outside the home. A total of thirty-seven outdoor uses in addition to lighting were counted. Different kinds of farming, however, required various amounts of electricity, with dairy and poultry farming using the most. The Agricola Dairy Farm near Gadsden used 13,500 kilowatt-hours per year, while forty other dairies averaged 2,000 kilowatt-hours. The largest poultry farm used 4,320 kilowatt-hours with the average consumption per poultry farm being 2,640 kilowatt-hours per year.[19]

A crucial observation was the small use of electricity on cotton farms. The planting, cultivation and picking of cotton depended on animal power or tractors and hand labor. Pumping water for irrigation was one of the few ways that a cotton farmer might require large amounts of electricity, but irrigation was not practiced widely in the South.

The seventy-four cotton farms averaged only 240 kilowatt-hours per year, and nearly all of it went for home appliances. Some 600 to 1,000 kilowatt-hours were considered necessary to make electrification economically feasible for both the farmer and the utility company. Lack of consumption was attributed to the weaknesses of one-crop farming with its low family incomes and low usage of power in growing cotton. E. C. Easter, chief field officer of the project, believed the final solution for Alabama and the South would be "the adoption of a more balanced program of agriculture that will make possible continuous use of electric power throughout the different seasons of the year."[20] M. J. Funchess, dean of the agricultural college at Alabama Polytechnic Institute, observed that "it probably will not be easy to develop profitable rural electric lines in those communities that grow little else but cotton."[21]

The Alabama project thus provided observations on more than the mechanical uses of power. It demonstrated that one-crop farming was an obstacle to electric development and, since electricity could be profitably used on other types of farms, it could also promote diversification. In this sense the project had shown that abundant energy could be a vital factor in the economic development of the area. The Tennessee Valley Authority (TVA) would confirm this supposition ten years later.

Despite its contributions toward an understanding of the farm use of electricity, the Alabama project failed to come to grips with the cost of service. In this respect it was no worse than other state CREA's. The Alabama Committee, if anything, made a greater commitment and contribution than any of the others. But cost was the primary obstacle, and the Alabama CREA, although admitting rates were a handicap, sought a solution through sales and education of the farmer, thereby tackling only half of the problem. Records of the Alabama Public Service Commission show that the project's rates were graduated, but still too high. In areas with five customers or less per mile—the typical rural population in the United States—the project's rates varied from 3.9 to 12 cents per kilowatt-hour, still above what the farmer could afford. To concentrate on finding new ways of using electricity while not revaluating costs prevented a breakthrough, a fact made clear when the 1930 Census showed only 2.7 percent of Alabama farms with service. Farmers still complained that high rates precluded the use of electricity.[22]

But the projects at Red Wing and in Alabama were more extensive than the others, demonstrating even more conclusively the power industry's weak commitment. The Virginia CREA extended power to forty selected families in 1924, but otherwise made no progress; the group admitted failure in 1929 and tried to reorganize, but it came to nothing. Most state committees were even less active.[23]

By 1930 CREA was regarded as a failure. The Census Bureau reported 14.3 percent of farms in the United States had electricity, but that figure was inflated since it included home generating plants; only 9.5 percent had central station service. Even though power industry spokesmen boasted that 418,453 farms received service, most were "suburban farms."

CREA belatedly dissolved itself in 1939, an empty gesture since it had been little more than a "window-front" since the end of the Alabama project. CREA never had any real driving force. "It was a dress sword," concluded one observer, "rather elaborately decorated, not much more hurtful than a hatpin in a free fight."[24]

In one sense, however, CREA was important. Through its sponsorship more than 200 uses of electricity were developed and improved. A large number of reports, bulletins, trade articles and pamphlets were published and kept farmers and electrical engineers informed of advancements. Most important was the demonstration of electricity's impact on rural life: sanitation was improved, diets were improved and more time was available for leisure and recreation. Best of all, the rural inhabitant, in the words of one utility officer, "had an even chance with his city brother in the comforts of life."[25] The Alabama study also clearly indicated electricity's potential as a catalyst in the diversification of agriculture.

But the CREA experiment ended with a question mark. While it improved knowledge of farm uses of energy, it left the high cost of serving the farmer in need of a solution. Even the Alabama Power Company, which made the most serious effort of all CREA participants, could not find a way to reduce its rates measurably even after 1925 when the company received 160,000 horsepower at low cost from the Wilson dam of the Muscle Shoals project. National Master of the Grange Louis Taber ably described the situation: "The problem of lower rates will continue until adjustments are made and there is a clearer understanding of the problems underlying the cost of manufacturing and transmitting electrical energy."[26]

Public power enthusiasts accused the industry of ignoring the farmer or making unreasonable demands for service. They also refused to accept the industry's figures on cost. For evidence they cited a study by the Wisconsin Electric Association, a private trade guild, that quoted line cost at $1,225 per mile, considerably less than the figure normally cited by power companies. Arrogance of utility executives and their lack of social conscience were responsible, critics insisted, for the poor state of rural service.[27]

Power companies were justified, however, in not serving some farms, especially those in areas where distances between households were considerable or in mountainous regions where construction costs were high.

Many farms were also submarginal and needed to be relocated on more productive land. For that portion of rural society, as high as one million farms by one estimate, the power industry could not be expected to provide service.[28]

Electrification also required a hefty initial investment by the consumer. Each home had to be wired, an expense averaging $100 at the time. Several appliances, plus lighting fixtures, were necessary, and expenditures for these items strained the pocketbook of the farmer; some needed small consumer loans. It took several years for families with few appliances and little need for electricity in their farming operations to build enough "load" to be profitable for the power company, and company executives understandably regarded these obstacles as too severe to warrant investment in the rural market.

Cost was primarily responsible, therefore, for the lack of rural service. Each party, the power company or the consumer, could not recover its initial investment soon enough. By 1930 it was clear that farmers were not likely to receive service within a reasonable time through the aegis of private enterprise. At the industry's rate of extending service, electrification of the remaining 5,000,000 farms would take 100 years. It became increasingly clear that an alternative was needed.

2
THE ALTERNATIVES: COOPERATIVES AND PUBLIC POWER

Because of the power industry's slow progress in serving farms, public power spokesmen such as Senator George Norris of Nebraska included rural service in their call for federal development of electrical energy. Reinforcing their belief was the successful operation of a limited number of electric cooperatives (co-ops) organized by farmers when CREA was in operation. Critical details of cost and operation were involved in comparing co-ops with utility service, but the successful use of the co-op on a broad scale in Europe and Canada also pointed to its feasibility. This device, however, violated the sanctity of private enterprise, making rural electrification part of the public power question.

Chroniclers dispute the date when the first electric cooperative was formed in the United States, but they agree that prior to World War I a few existed. The Stoney Run Light and Power Company at Granite Falls, Minnesota, was established in 1914, and the Kegonsa Electric Company at Stoughton, Wisconsin, was formed in 1916. In addition to rural customers the latter served 200 summer cottages at Lake Kegonsa, indicating that it was not purely a farmer-owned cooperative. After World War I, co-ops increased as the need and demand for service increased. Most were concentrated in the Midwest and Northwest; Iowa ranked first with twelve, followed by Washington, Minnesota and Wisconsin, respectively. Only two southern states, North Carolina and Virginia, had co-ops, each with two. In some cases farmers organized and built lines on a cooperative basis and turned them over to the local power company to operate

and maintain. Only those independent of power companies were right-
fully considered co-ops, however, and about fifty of these were in operation by the time REA was started in 1935.[1]

Establishing cooperatives held some promise for solving the problem
of electrification. Typically membership ranged from fifty to 200 families. Miles of line and the per mile density of the co-ops also varied:
one had eighty-two members with nineteen miles of line. Initial construction costs were met by a membership fee normally falling between $100
and $200. After the first payment of the membership fee, about 25 per-
cent, farmers usually paid the balance through installments incorporated
into the monthly bill. Power companies by comparison required total
payment for the line at $2,000 per mile, or else demanded a cash deposit
seldom less than $500. Co-op rates, too, were comparatively low: one
sold at 2.7 cents per kilowatt-hour, while another charged 3.3 cents per
kilowatt-hour. In some instances, however, rates were as high as 9 cents,
but the generally lower price for energy sold through cooperatives strongly
suggested that power company rates, usually 8 to 12 cents per kilowatt-
hour, were inflated. Most of the co-ops were able to charge less through
the use of graduated rates in which the first 100 kilowatt-hours cost 4 to
10 cents each, but with each succeeding block of power consumed, the
price went down until the recipient paid at a reasonable and affordable
rate.[2]

Despite lower rates and smaller initial costs, most cooperatives were
solvent. The Parkland Light and Power Company at Parkland, Washington,
founded in 1914 with only twelve members, had by 1935 300 members
and a paid-in capital of $20,806 with a reserve fund of $10,000. Surplus
earnings were returned to the members through lower rates, or new ex-
tensions and street lights in Parkland. The Parkland co-op, having some
of the lowest rates, was considered successful "both from the financial
and social point of view."[3] The Ferry Light and Power Company at
Burley, Idaho, was organized in 1919 with eighty-two members and was
still in operation when REA was created. It had paid-in capital of $12,625
and a reserve fund of $1,000. Like other co-ops, Ferry Light and Power
returned surplus earnings or profits to its members through lowered rates.
Expenses were reduced by letting members read their own meters and
send in their payments, a practice later adopted by REA. Whenever
possible members also provided their own maintenance, but the more
sophisticated repairs required trained linemen.

The mortality rate of electric cooperatives, however, should not be overlooked. Too often they were spontaneously organized without a legal counsel, with little technical advice on construction, with little managerial direction and with inadequate financing. Even Iowa, which claimed the highest number of co-ops, had several failures, mostly due to poor management and careless maintenance of the lines. In one case a pole on a sloppily constructed line blew down in a storm, killing a child. Lack of coordinated service for rich and poor farms was another fault; prosperous farms in densely settled areas were preferred over the more widely scattered homes, resulting in the establishment of associations too small in membership for successful operation. Area coverage, the serving of rich and poor farms together, would have produced larger cooperatives better able to retain properly trained personnel.[4]

Cooperative associations suffered not only from lack of qualified management and leadership but from attacks by the electric companies. Co-ops were at the mercy of the companies because the latter supplied the energy distributed along the lines, and if a company were so inclined, it could charge rates too high to let the association continue operation. If such were the case, the co-op was unable to save funds for maintenance, depreciation, insurance and reserves. Eventually the consumers willingly gave up ownership to rid themselves of the responsibility. Power companies sometimes acquired the more lucrative lines of a co-op. The Morrison's Cove Light and Power Association of Pennsylvania was launched in 1925 after the Penn Central Light and Power Company refused to serve farmers living within its franchise. Once the co-op was successfully established, however, Penn Central acquired control, offering the original stockholders a 120 percent profit. But Penn Central refused to take the less profitable lines which were unable to stand alone, thus causing recipients along those lines to lose service. More cooperation and less desire for personal gain on the part of the members living on the lucrative lines would have meant an equitable benefit for all concerned. Electric cooperatives were also prohibited by law in some states, while in others the public service commissions were unfriendly to them since such commissions usually were instruments of the power industry. "Neither the private utilities nor the unfriendly commissions," wrote one REA administrator, "ever seemed to tire of taking pot shots at the farm electric cooperatives."[5]

Except for success in some local areas, progress toward electrification moved no faster via the self-help concept than through the power companies. Reasons for the poor showing included opposition from industry, improper management by the farmers themselves and technical and financial barriers unsurmountable without forceful and expert assistance. Still, the cooperative offered real hope: it was like a yet-to-be-discovered cure for a disease; if proper management, technical knowledge, special financing and legal clearance were available, farmers, so it was thought, could and would organize and solve the problem themselves.[6]

This last thought was reinforced by the presence of cooperatives in Canada and Europe where in some countries 90 percent of the farms had service compared to only 10 percent in the United States. Conditions essential to the success of cooperatives varied widely in Europe and did not always furnish a pattern applicable to the United States. Public power proponents, nonetheless, pointed to the progress in cooperative distribution of electricity overseas as an indication of the possible advancements in rural service in the United States through public ownership. Sweden had one of the best-known such developments, thanks to a study by Marquis Childs, a writer for the *St. Louis Post-Dispatch*. He discovered that "rural electrification in Sweden is accomplished almost entirely through cooperative societies made up of consumers of electric power."[7] Farmers organized and raised the capital for the co-op through membership fees, bought the energy from the state, redistributed it among themselves and maintained their own lines. They had to follow certain guidelines set down by the Royal Board of Waterfalls, mostly construction of lines to ensure area coverage and rate approvals. Rates averaged 3 cents per kilowatt-hour, and by 1936, 50 percent of Swedish farms were electrified.[8]

France, with 50 percent rural population, had achieved 71 percent electrification by 1930 by means of rather heavy subsidies. The French government also furnished cooperative members with engineering and technical advice, subsidies for construction and loans at low interest rates. Educational programs to acquaint farmers with the value and use of electricity were provided. Maximum subsidy in France was 50 percent of the cost, although it was higher in a few cases. Germany depended less on subsidy, although prior to World War I indirect aid came in the form of cheap credit. The real push came after the war: the Weimar Republic claimed 60 percent rural electrification in 1927. German cooperatives purchased energy from public and private sources and distributed it to members. Other countries far ahead of the United States included Finland with 40 percent rural electrification; Denmark with 50 percent; Czechoslovakia with 70 percent and New Zealand with 35 percent.[9]

Several factors accounted for the prominence of European cooperatives. World War I had drained the supplies of fuel oil and stimulated the need to resort to electricity; agriculture flourished for several years after the war, enabling farmers to modernize their homes; and the European countryside had a higher density of population per mile than the United States providing greater revenue from rural lines. European governments also felt a social responsibility to extend power into remote areas. An important characteristic of foreign electrification, therefore, was public ownership. Private interests in the United States argued that the use of subsidies accounted for European advancements, but their opponents attributed the American lag to greed and social irresponsibility of the power industry.[10]

An example of rural public service closer to home was the Hydro-Electric Commission of Ontario, created in 1908. The Commission generated and sold energy to municipalities throughout the province, whereupon it was redistributed to consumers through locally-owned systems. Half the cost of the distribution lines was accepted by the provincial government. With this subsidy, electrification developed rapidly; in the southern portion of the province where the bulk of the rural people lived, 27 percent of the farms had service. A graduated rate schedule was followed, beginning with a charge of 6 cents per kilowatt-hour for the first block of energy and dropping to 2 cents per kilowatt-hour for the last. Hence the average rate was about 3 to 4 cents. The Commission also provided small loans to rural customers for the purchase of appliances. This service enabled residents with small income to build up their use of energy rapidly, thus keeping rates low.[11] Senator Norris saw the Ontario project as "a most wonderful demonstration of the possibilities for the generation and distribution of electric current [that] had been given to the civilized world. . . ."[12]

Ontario lent itself very well to Norris's fight for federal development of the Muscle Shoals project on the Tennessee River. He believed Muscle Shoals could be the American counterpart of the Canadian system and that it presented an opportunity to gauge what he considered the real cost of electric service for rural or urban markets. While the potential generation of low-cost energy for farm service was not the only reason he wanted development of Muscle Shoals, it was one of the most important.

Norris's first step in that direction came in 1922 when he submitted his bill for government operation of the federal properties at Muscle Shoals to counter Henry Ford's offer to purchase the entire project for five million dollars. Ford had proposed to feed the hydroelectricity generated at the dams strictly to manufacturers, thus eliminating any possible agricultural use for it. The senator recommended, however, that the hydroelectric power be distributed within a 300-mile radius with preference given to states, counties and municipalities, thereby benefiting farmers and small town users of electricity.[13]

During the summer of 1925, Norris toured the Province of Ontario to get a first-hand look at a public power project operating on a large scale, and that was when he fully realized the tremendous opportunity to improve rural life with electricity. His trip was prompted by the controversy over the "Wyer pamphlet," a strongly worded and widely circulated attack on the Ontario system by an agent of the utility industry. The pamphlet maintained that public power, namely Muscle Shoals, was not in the best interests of the United States. Norris was already well informed on the Canadian operation, and in a special session of Congress in March 1925, he discussed the allegations made in the pamphlet. Pointing to the advances made in Ontario and inserting into the *Congressional Record* a variety of material that described the improved farm life, Norris was able to refute the Wyer pamphlet and at the same time focus attention on the possibility of public electrification in the United States.[14]

It was during his trip to Ontario three months later that Norris saw the real meaning of the Ontario system. He reported that he saw the end of much of the drudgery of farm living, that he saw a sense of pride, personal satisfaction and happiness in the rural home, and that he noticed less desire on the part of youth to forsake the country for the city. The senator went home even more convinced the Canadian project was a fair indication of the superior efficiency of public power, and that it demonstrated how higher rates in the United States were the result of alleged greed for profits by the privately-owned electric companies. "Ontario Hydro," wrote his biographer, "thus served as an important weapon in Norris' arsenal of information," and he "constantly referred to Ontario Hydro in his war against utility companies."[15]

Indicative of the influence of the Ontario system was the passage of legislation by the State of Washington in 1930 authorizing the creation of Public Utility Districts (PUD) to serve rural areas. Since 1919 the state Grange had campaigned for public service, pointing to Ontario as a successful example. After a long battle against the power companies, the Grange, assisted by public power proponents such as Homer T. Bone, persuaded the legislature to enact Initiative No. 1 authorizing PUDs. Progress in serving farms was slow, however, and Washington was the only state to establish such a program. A similar effort failed in Oregon. Few states, furthermore, had comparable resources for hydroelectric development.[16]

Included among the small band of early crusaders for public service was Judson King, long-time director of the National Popular Government League (NPGL). Once described as "a giant among pigmies, a lion set down among rabbits," he belonged to the camp of social planners who believed in public control of America's natural resources.[17] Like Norris, he saw in Muscle Shoals and similar sites the potential for inexpensive hydroelectric power for the average citizen's use.

Born in Waterford, Pennsylvania, in 1872, King had a long and stormy career as a propagandist, up to the moment he died in 1958. Orphaned at the age of six, he grew up on a farm and left at seventeen to attend Battle College and later the University of Michigan. In 1902 he founded the *Denison* (Texas) *Morning Sun*; three years later he went to Toledo, Ohio, as editor of the *Independent Voter* and worked with Mayor "Golden Rule" Jones. It was here, he later wrote, that "my own active acquaintance with the aims and political tactics of the electric and gas industries had begun. . . ."[18] He became a lecturer for NPGL in 1910 and then its director in 1913, a post he occupied until his death. Best remembered for his support and collaboration with Norris in the fight over Muscle Shoals, his career in public causes continued after TVA was established. From 1935 through 1944, he was special consultant to REA. But his real contribution came as director of the NPGL, because it gave him the opportunity to supply critical information for others like Norris to air. Their relationship was very close, and "for many years," wrote a friend, "Judson and his wife drove Senator Norris . . . to his office . . . often taking the Senator home again in the evening."[19]

King wrote the *National Popular Government League Bulletin,* one of the best sources of information on public power. He was particularly effective at furnishing technical and cost data to refute private interests

and at exposing agents of the power industry. In this last regard he exposed college professors who purported to offer independent and objective testimony on the merits of public and private power while on retainer to power company interests.[20]

One such example was E. A. Stewart, professor of agricultural engineering at the University of Minnesota, who in 1926 wrote a pamphlet widely circulated in electrical and political circles and appearing to be an impartial, scientific study of Ontario rural electrification. According to Stewart, the cost of service in Ontario was roughly the same as that in the United States. Public power allegedly did not hold promise for lower rural rates. Copies were sent to members of the 60th Congress to coincide with Norris's attempt to prevent the lease of the Muscle Shoals plant to private interests. Through King's footwork, it was demonstrated that NELA had paid Stewart to write the pamphlet. Officers of the Ontario project, after King informed them of the situation, reported that Stewart misled them and acted without authority in his investigation and in writing his report. Stewart lost his credibility and the University of Minnesota severed its relation with him. Stewart then went to work for a power company. "The Stewart case," King wrote later, "became an ugly example of how university men were used in utility propaganda."[21]

Proponents of public power espoused the common attitude that disparities in country and city life were responsible for the migration into towns and cities. Few could bring the hardship of rural life into the issue like Norris. He was a product of Nebraska, and "from boyhood," he once said, "I have seen first-hand the grim drudgery and grind which had been the common lot of eight generations of farm women, seeking happiness and contentment on the soil." He eloquently spoke of "the heat of those summer days in a farm kitchen . . . where humidity and the blazing sun combined with the wood-burning stove to create unbearable temperatures." He could describe the "thousands of women . . . growing old prematurely, dying before their time, conscious of the great gap between their lives and the lives of those whom the accident of birth or choice placed in the towns or cities." Inherent, therefore, in Norris's struggles, ranging from public power to inefficiency in government, was relief and assistance for the farmer. That was why he regarded electrification as one of the most important means to improve rural life.[22]

Difficulties in establishing cooperatives, plus differences of opinion over federal development, demonstrated that the electrification of agriculture by public means was also a problem of immense proportions. The speed at which farmers received service through cooperatives was pathetically slow because of the managerial and financial problems involved. Programs similar to the one in Ontario required broad support, and in the conservative climate of the time, such support was unavailable. Limited success of the cooperatives, however, left the haunting feeling that if enough expertise and financial backing were available to farmers, they could organize co-ops and serve themselves. The problems of extending service to farmers, whether by private or public means, had such magnitude that only a particularly well-conceived plan of great proportions could provide a solution. The search for that solution was already underway in Pennsylvania.

3
MORRIS L. COOKE
AND GIANT POWER

Even though Senator George Norris was widely regarded as the champion of public power, he had to share that honor in so far as rural electrification was concerned with Morris L. Cooke. It was Cooke who put together the administrative, technical and operational jigsaw that became REA in 1935. His involvement with rural service began ironically in 1923 when CREA was founded, but his approach to the problem differed greatly with that of the power industry. Through his service with Gifford Pinchot in Pennsylvania, Cooke designed a broad-scale energy development plan known as Giant Power. Simultaneously, however, the power industry was developing an energy plan called SuperPower that was not related to CREA. These two concepts incorporated more than rural service, but they epitomized the divergent views toward the subject. Giant Power and SuperPower, plus CREA, cooperatives and the Ontario plan demonstrated the extent of the search for a feasible method of serving farms prior to the creation of REA in 1935. Cooke's work in Pennsylvania, however, came closest to serving as the precedent of REA.

Born in 1872 to old-line Philadelphians, Cooke trained as a mechanical engineer and established a consulting firm in his home town. A highly praised engineer and an avid follower of Frederick Taylor's principles of scientific management, he received an appointment in 1912 as director of the Philadelphia Department of Public Works. It badly needed reorganizing, but most of all it needed to be freed of special interests. Cooke engaged in several battles with the city's private contractors, but the most dramatic incident occurred when he filed suit against the Philadelphia Electric Company (PEC) for a reduction of the city's rates.[1]

His step was quite unusual because utility companies in most cities had generally been allowed to set rates without challenge. Cooke knew little about electrical engineering, and with his small staff combatting one of the country's largest power companies, the suit was cast as a David versus Goliath battle. Surprisingly Cooke won and became known as an innovator in his profession and an economic reformer in general. During this experience Cooke learned that state regulatory agencies were often tied to the power companies they were supposed to oversee, or that they lacked anyone competent and willing to present anything other than the view of the private interests.[2]

In 1923 Cooke's expertise and reputation received the attention of Pennsylvania's newly elected governor, Gifford Pinchot. Known for his efforts in conservation when he served as advisor to President Theodore Roosevelt, Pinchot was determined to inject the old progressive spirit and zeal into Pennsylvania state affairs. He wanted to extend state authority in regulating capital and business and also promote social and economic welfare. In regard to the power companies, Pinchot saw them as irresponsible in setting rates and in failing to extend service into the countryside. After explaining his goals to Cooke, the new governor persuaded him to serve as director of the specially created Giant Power Board. For Cooke it was another opportunity to practice his social-minded philosophy. "I am a public engineer," he wrote a friend, "and not a private engineer."[3]

Determined to develop Pennsylvania's electrical resources to minister to the needs of society, Pinchot and Cooke assumed that whatever could be accomplished in the Commonwealth could also be extended to the federal level. Their special concern for the development of rural electrification was made clear when Pinchot addressed the 1924 meeting of the National Electric Light Association whose offspring, the CREA, was half-heartedly trying to increase the number of electrified farms. "We are endeavoring to bring power . . . to every man's door at standard prices. . . . The ultimate system which we should have . . . must transcend state lines and is likely to become nationwide. . . . One of our chief ends is the supply of electric service to the farms" because without the benefits of modern technology, farmers had been isolated "and a decay of rural life resulted."[4]

Indicative of the importance placed on rural service was the composition of Cooke's staff. Three members were retained for this one aspect of the study: George Morse, who had been with Cooke in the PEC fight, handled much of the field work; Ralph U. Blasingame, a professor of agricultural engineering at Pennsylvania State College and author of several books on the farm uses of electricity, directed research; and Philip P. Wells, longtime associate of Pinchot in the Reclamation Service and now deputy attorney general of Pennsylvania, used his authority to obtain less accessible information on rural rates, generation, construction and other aspects relating to cost. Wells was a particularly effective member of the team because he could use the investigative powers of his office to obtain critical data on costs that power companies would not have released voluntarily.[5]

One characteristic common to all members of the staff was their public point of view, a sharp contrast with the view of CREA. To enable all residents of the state, rural and urban, to achieve the benefits of electrical technology at lower cost, the Giant Power Board was convinced that a dramatic, new approach to the generation and use of power was necessary, one that encompassed each level of operations. The Board, therefore, analyzed every aspect of energy operations in Pennsylvania, including farm service, city gas supply, water power development, mine-mouth electric generating plants, public utility regulation, high voltage transmission and related matters. Energy use was considered from the standpoint of all interests—the large and small consumer, the power industry and the government, but always with a careful regard for future development and in light of human concerns and efficient use of resources. The achievement of any goal, especially the electrification of agriculture, depended on a fully integrated system at each level of operations. Use of power on a farm, for example, would depend on the efficiency of coal mining operations across the state. In point of view and in its approach to the study of farm service, the Giant Power Board was the antithesis of the power industry's CREA that focused on increased sales.[6]

Cooke's staff went into other states to make comparative studies of rural service. Morse thoroughly analyzed the use of electricity in the farm community of Waukesha, Wisconsin, served by the Milwaukee Electric Railway and Light Company. He also studied the use of energy in Toronto Township, Ontario. In each case it was clear that electrification was economically feasible and that the families having service enjoyed a far more rewarding and comfortable life. Average cost of service in Waukesha was 5 cents per kilowatt-hour, and the recipients claimed that the average monthly bill, about $7.87, was repaid in time, efficiency and comfort. By comparison Pennsylvania farms with central station service benefited in a similar fashion, though rates were higher, about 8.2 cents per kilowatt-hour. Only 6 percent of Pennsylvania farmers had service, however, and Morse in his final report predicted that "when farmers in Pennsylvania wake up to the fact that electricity can transform their lives from drudgery and ineffectiveness to comfort and accomplishment, nothing will prevent them from having it."[7]

Study of rural rates, always a thorny question, was as much a problem for the Giant Power Board as for anyone else. At one point Cooke sharply disagreed with his staff that rural rates could never fall below urban rates. He thought that under certain conditions rural service might be cheaper. Opportunities for consumption of energy, he anticipated, were greater on farms than in city homes, offering a real chance to achieve lower cost per kilowatt-hour. Cooke withheld judgment, however, until electrification was widespread enough to prove or disprove his point. But he was not satisfied that "urban rates have been determined beyond peradventure of error," and since rural rates nearly always consisted of the nearest urban rate plus a surcharge, he refused to believe that farm service had to be higher. With tongue in cheek he reminded Wells that "a large part of my job in connection with Giant Power is holding views which all my associates consider irrational or worse, so that to have only four against me on a point is not a bit discouraging."[8]

The Giant Power Board submitted its report in February 1925 with proposals for legislation that would implement its recommendations. In his introductory message transmitting the report, Pinchot described Giant Power as "a great pool of power into which power from all sources will be poured . . . and the chief idea behind it is not profit but public welfare."[9] To reach that end, the Board proposed a delicately balanced plan of full-scale regulation and comprehensive development of energy resources, going beyond Pennsylvania borders if necessary. Each step in the plan was interrelated with the next, thus, extension of rural service would depend on the efficient operation of each facet of the plan.

The legislature would have to create a new Giant Power Board with broad regulatory powers. To reduce the cost of electricity, the proposed Board would require new generating plants to be located at coal mine sites and require them to have by-product recovery, the recycling of coal waste. Current of no less than 110,000 volts would be carried over cross-country power lines to the cities and rural areas, eliminating the expensive rail transport of coal to the generating plants. The Board would also select transmission routes and all lines were to be interconnected. Negotiation with other states would be conducted for the acquisition and disposal of energy, and Congress would have to approve such arrangements for interstate transport of energy.[10]

The state Public Service Commission would be given greater powers— it would have complete control and supervision over all contracts for construction, lease, consolidation and financing of electric companies. It would also control the issuing of securities. All electric company stock would have a standard valuation of $1.00 per share, and each company operating in the state had to have a new valuation as of January 1, 1926. The purpose for the latter provision was to standardize valuations in order to have a basis for setting standard rates, one of the original and most important goals of the Board. In other words valuation and rates would be based on the theory of "prudent investment" and not "reproduction cost" theory as practiced by the industry.[11]

Development of the rural market depended on the proposed synchronization of energy operations throughout the state. It also depended on strong regulation. With more efficient coal mining, high-voltage transmission, standard valuations of stock and the other features of the plan, the cost of distributing electricity to the farmer would, it was thought, fall to a level profitable for the power companies and still cheap enough for the consumer. The state's Public Service Commission would be given authority to make power companies justify why they should not serve particular areas. Cooke also proposed practice of area coverage, blending poor and rich farms together so that revenue to the power company from the latter would offset the losses of the other. Poor farms after an initial waiting period were expected to use enough energy to be self-supporting. By following such a plan, all houses could be served. Realizing that utility companies might still need prodding, the Giant Power plan recommended legislation giving farmers the option of establishing cooperatives. Farmers could exercise that option, however, only when companies refused to serve them. Cooperatives were thus viewed as an alternative; the real impetus for extension of rural lines, it was hoped, would come from private enterprise.[12]

It was characteristic of Cooke to look to the electrical industry as the supplier of energy to the farmer. He was convinced that since the power companies already had the resources, they were best suited to serve the rural market; it would be too expensive and time-consuming to handle the situation by other means. He would not abandon the cooperative, however, for he wanted an alternate means available if his predictions proved wrong. Later as the first administrator of REA he had a chance to test his idea.

The proposals were bold and, as the Board admitted, required a new definition of some legal rights of utility companies. Proponents were firmly convinced, nonetheless, that the intricate and complex world of electricity merited an altogether new approach to regulation, especially since 80 to 90 percent of the industry in Pennsylvania was in the hands of holding companies. Winning passage of the plan, it was realized, depended on overcoming the vested interests.[13]

The latter were determined to prevent enactment of the Board's recommendations. Opponents included the Pennsylvania Electric Association, the National Electric Light Association and the Investment Bankers Association, but in a broader sense the opposition amounted to the community of private enterprise. Giant Power was in sharp disagreement with SuperPower, the electrical industry's own plan to bring about a large-scale interconnecting system of energy somewhat along the lines of the Pennsylvania proposal. These two concepts symbolized the public and private approach to the development of energy resources at a time when, wrote one historian, "electric power represented the potential key to a new kind of civilization."[14]

The concept of SuperPower originated during World War I when many utilities pooled their resources to meet energy demands. Shortly after the war, the United States Geological Survey team, headed by consulting engineer M. S. Murray, studied electrical resources in the country. The survey team along with industry saw the need not only to develop hydro and fossil-fuel sources of energy but to let them complement one another. They recognized, too, that more generating plants had to be built and connected with cross-country transmission lines in order to pool power on a grand scale. Such an achievement would mean greater efficiency at production points and alleviate shortages of energy, enabling American industry to abandon the use of expensive and inefficient steam power. SuperPower included no plans for the by-product recovery of coal, but suggested that generating plants might be built near the coastline so as to harness the energy of the tides. Obviously, use of tidal energy called for technological research and massive capitalization, two critical barriers for which the SuperPower planners had no solution. SuperPower also recognized the acquisition and dispersal of electricity across state lines and recommended federal and state cooperation in regulating interstate use.[15]

In spite of many similarities, the two plans had real differences. Super-Power had no proposals for regulation except to leave it in the hands of the state public service commissions with no change. Revaluations of stock and property as proposed in Giant Power were never entertained, which according to the latter were critical if rates were to be fair. Super-Power would also permit holding companies to remain. The cost of service to the consumer, furthermore, did not enter into the thinking of Super-Power proponents, except in the sense that electricity was cheaper than steam. Rural electrification was recognized as desirable, but secondary to the energy needs of industry, and nothing was proposed, therefore, to encourage energy use in the rural sector. The plan only sought to supply industry with electricity on a larger scale than before by taking advantage of new technological developments. SuperPower did not deal with the all-important matter of rates, nor did it propose extending service to the small consumer. That is why Pinchot referred to the principal object of the plan as "profit for the companies." He described Giant Power, on the other hand, as "a plan to bring cheaper and better electric service to all those who have it now, and to bring good and cheap electric service to those who are still without it."[16]

If Pinchot was the spokesman for Giant Power, his counterpart for SuperPower was Secretary of Commerce Herbert Hoover. Cooke had met Hoover during World War I and had high respect for him, even to the point of participating in an effort to get the Republican presidential nomination for him in 1920. Philosophically the two agreed on the importance of large-scale electrical development. "Every time we cheapen power and decentralize its production," Hoover stated, "we create new uses and we add security to production; we also increase the production; we eliminate waste; we decrease the burden of physical effort upon men. In sum, we increase the standards of living and comfort of all our people."[17] Cooke had no quarrel with such statements, but each man still represented a different school of thought on resource development. Hoover wanted to keep the federal government out of the power business and persistently said that "the regulation of our utility rates to protect the public is the responsibility of our state government."[18]

Nowhere was their disagreement more evident than in the matter of rural electrification. For Cooke the most promising part of the Giant Power concept was "the distribution of current to the rural population together with a reduction of rates—these are the immediate and definite objectives of Giant Power."[19] Hoover saw "the agricultural problem as one of first getting our primary system onto right lines," meaning that farm service must wait until SuperPower industrial development was achieved.[20] Cooke was convinced that 50 to 75 percent of America's farms could be electrified within ten to twenty years if the Giant Power Board's recommendations were implemented, but at the rate the electrical industry was progressing, he explained, it would take 100 years to connect America's farms.

Giant Power and SuperPower became the focal point of the debate over rural electric development. On Cooke's side were "progressives" who, like Pinchot, wanted to assert public authority in a wide range of affairs besides power: labor, agriculture, conservation and consumer rights. Men as influential as Samuel Gompers, George Norris, Henry L. Stimson, Alfred E. Smith and Sir Adam Beals of Ontario were part of the Giant Power arsenal. But SuperPower was the stronger combatant. It enlisted the executive branch of the federal government, and it had the support of the banking and mining industries, the national Chamber of Commerce and, of course, the National Electric Light Association. The press lined up behind SuperPower, although Cooke received kind treatment from some publishing houses and newspapers.[21]

The hearings over Giant Power in the Pennsylvania statehouse demonstrated the extent of the resistance which Cooke and his associates faced. Representatives of the electric companies in the state and the Investment Bankers Association were the most polemical. Mainly they argued that the engineering was unsound, but their real target was the proposed stiffer regulatory powers that would be given to the Pennsylvania Public Service Commission. Provisions to authorize the creation of cooperatives recommended that first preference be given to the companies serving rural areas. Opponents insisted that use of cooperatives amounted to government ownership and control of an area of the market designated by law to be the prerogative of private enterprise. The president of the Pennsylvania

Electric Association saw no need to encourage cooperatives because "rural electrification is making real progress at the present time."[22]

Academicians, allegedly above special interest influence and considered capable of making an accurate analysis of the power situation in Pennsylvania, joined in the chorus of protest against Giant Power. A. E. Kennedy, professor of electrical engineering at Harvard University, testified that "the United States has led and leads all countries in the world in the distribution of power and light to homes . . ." although only about 5 percent of American farms had service at the time when some countries had already reached the 50 percent mark. If Giant Power passed, the Harvard professor continued, "the people on the farms will be unable to get the power they need . . . the present system is working well . . ." and the proposal to establish electric cooperatives "was a most dangerous plan."[23] Judson King later proved that Kennedy at the time of his testimony was on retainer from the National Electric Light Association. Paul M. Lincoln, director of the Cornell School of Engineering, insisted the rural electric plans of Giant Power were not technically feasible. He was embarrassed, however, when deputy attorney Wells demonstrated that on another occasion the Cornell professor had fully endorsed the same kind of plan when set forth by a private company. The professor lamely extricated himself by saying that he had "forgotten" about the earlier report. As Senator Norris found in his struggle over Muscle Shoals, Cooke learned that academicians were generally no friend of the dirt farmer.[24]

Passage of the Giant Power plan was hopeless. Opposition was too stiff, consisting of power company executives, vested interests, a majority of the legislature and a majority of the joint committee of the Senate and House which conducted the hearings. For the conservative climate of the times, the Giant Power plan was too bold, and the legislature voted it down.

Giant Power was important in the development of rural service, because it was the first study of the problem from every angle, including rates. CREA, too, was investigating farm service, compiling an impressive array of engineering data, but CREA never dealt with cost, the ultimate barrier. Thanks to the Giant Power study, it was recognized that farmers were potentially greater users of energy than city residents and would consume an amount sufficient to offset the higher cost of serving them. The study also showed that if rural rates followed a graduated scale, residents would tend to use more energy; the Giant Power Board agreed with the power industry, however, that a monthly minimum fee, about $1.00, should be charged. As long as there were at least three farms per mile, rural service could be self-supporting, but special treatment or a subsidy might be required for areas with lower populations. When Cooke became the first administrator of REA, his policies were based on these observations, and some of the Giant Power personnel became the skeleton of REA staff.[25]

Cooke's experience as director of the Giant Power Board was a milestone in his career. He had studied the problems of rural service to the extent that few men could equal his knowledge of and dedication to the subject. For the millions of Americans needing modern home comforts, Cooke had a plan to help them. He became a leading proponent of electrification which was rapidly being recognized as one of the most potent forces for the improvement of farm life.

With the failure of the Giant Power plan, service to farms in Pennsylvania proceeded according to the wishes of the power companies. In 1926 the State Grange and State Council of Farm Organizations persuaded the Public Service Commission to issue Order 27 which instructed power companies to supply farms under certain conditions specified by the Commission. Order 27 was popular with rural inhabitants and partly acceptable to some of the members of the Giant Power Board. But the utilities protested and persuaded the Commission to issue Order 28, superseding the first and requiring the recipient of service to pay a portion of the cost of extension, which was the regular practice throughout the United States. To the new rule the *Philadelphia Patriot* replied that the farmer was only a "helpless foundling on the doorstep of the public utility corporation" because "what chance has he against the highly paid lawyers and experts of the power companies?"[26] Order 28 brought about the creation of the Joint Committee on Rural Electrification, consisting of nine farm representatives and seven from the companies whose job was to work out cooperative arrangements between farmers and the utilities. Some progress was made with the new arrangement, for within three years 11,832 farms were connected. But this progress was due to the willingness and ability of the particular farmers to pay construction costs. The real problem remained, and Philip Wells swore that "the farmers of Pennsylvania will get no general relief from this oppression until they compel their legislative representatives to enact the provisions of the giant power legislation."[27]

Cooke returned to his private practice in Philadelphia with no opportunity in sight to act on his proposals. Another turn came, however, with the election of Franklin D. Roosevelt as governor of New York in 1928.

The governor wanted to promote rural electrification, for "after all," he said, "our principal fight on the question of electrical power is on behalf of the household and farm consumer . . . for the rates to the very large consumers of industrial power are about as low as they can be."[28] Roosevelt had become acquainted with the problem during his residence at Warm Springs, Georgia, in 1924, when he received his first electric bill. The rate was 18 cents per kilowatt-hour, "about four times," he said, "what I pay at Hyde Park, New York. That started my long study of public utility charges for electric current and the whole subject of getting electricity into farm homes."[29]

The governor wanted a hydroelectric plant on the St. Lawrence River in order to increase the supply of energy that could go to the domestic and rural consumer. For that purpose he managed, after some difficulty with the legislature, to get a bill passed in 1931 creating the New York Power Authority. He appointed Cooke a trustee of the Authority.

This new appointment came at a critical time for Cooke. Shortly before he had suffered a setback at the hands of President Hoover, still Cooke's friend but opposed to his thinking on power. Recognizing the need for greater employment, Cooke suggested to Hoover in 1931 the possibility of a national plan for rural service not only because "electric service on American farms would be appreciated out of proportion to its intrinsic cost and add immeasurably to the effectiveness of farms" but it would also employ thousands of workers. Hoover's secretary sent a polite and brief reply, a "brushoff" as described by one writer. Cooke had tried with a slightly more detailed explanation of his proposal to Colonel Arthur H. Woods, chairman of the President's Commission on Employment. Cooke recommended "some emergency non-profit organization . . . in full cooperation with the electrical industry." The commission approved it, but Hoover killed it. It was apparent that until new leadership came into the White House, action on this crushing problem would not be forthcoming at the federal level.[30]

As a trustee of the New York Power Authority, Cooke had to devote most of his time to the plans for the St. Lawrence hydroelectric plant, but he continued to carry forward rural electrification, both from the standpoint of promoting public awareness of the problem and increasing his own understanding of the technical aspects. To Senator Robert M. LaFollette of Wisconsin, who advocated public works as a means toward fuller employment, he recommended a federal construction program. To

Senator Robert F. Wagner of New York he urged that one billion dollars be spent "to make life on the farm a wholly different thing." Cooke also assisted the Joint Committee in Pennsylvania, always complying with requests for guidance and data, and always supporting anyone trying to get electricity to farms. He also remained in close contact with Senator Norris in his fight for federal development of power at Muscle Shoals.[31]

Cooke's real contribution at this point, however, was the cost study of rural service made on behalf of the New York Power Authority. Utilities had always insisted that it cost about $2,000 per mile to build the lines. Critics of the industry, including Cooke, had never accepted that figure, and some insisted it was either a falsification of the records or based on inflated and outdated rate formulas. Studies of generating cost had been conducted, but no comprehensive study of rural distribution costs had been made. Industry critics had no scientific evidence to support their claims, and Cooke undertook to study distribution costs from the generating plant to the living room.[32]

Cooke and his staff itemized all the distribution costs that went into rural electrification, making allowances for every item of expense in the areas of construction, maintenance and management. They came up with a construction cost of $300 to $1,500 per mile. The impact of this conclusion was monumental; it meant that current could be extended at a fraction of the investment that had widely been accepted as standard to that time. Cooke regarded it as "a very important landmark in electrical development."[33]

Results of the study were presented in January 1933 before the Institute of Public Engineering sponsored by the New York Power Authority and several other state public service commissions. But electric companies did not hail the findings as the solution to a long unsolved problem and made no changes in their operations. Public power enthusiasts were convinced, nonetheless, that the principal barrier to farm service had been overcome. "Widespread rural electrification," Cooke told his fellow trustee Frank Walsh, "is socially and economically desirable and financially both sound and feasible."[34]

By the eve of Franklin D. Roosevelt's inauguration as president, the conviction grew that rural service was technically and economically feasible. Even the American Farm Bureau, which had belonged to the now defunct CREA, had a change of view. In 1932, Edward O'Neal, newly elected president of the Farm Bureau, informed Hoover that "the time has now come to regulate the power group . . . power should be clothed with public interest."[35] What the Farm Bureau had once regarded as blasphemy, it now considered essential to the general welfare.

Twice Cooke had stepped forward with a plan to help farmers acquire service, and twice he had been set back. Hoover was at best only instrumental in the first case, but was chiefly responsible in the other. Cooke sensed a new day during the presidential campaign of 1932 when Roosevelt set forth his views on power at Portland, Oregon. Discussing a wide range of power-related matters, the presidential aspirant touched Cooke's heart when he declared that electricity "is no longer a luxury. It is a definite necessity. . . . It can relieve the drudgery of the housewife and lift the great burden off the shoulders of the hardworking farmer. . . ."[36]

Roosevelt's remarks moved Cooke to write a column for the *Philadelphia Evening Ledger,* announcing his support of the Democratic candidate, even though he was a Republican. "Surely no public man in my generation has sensed the importance of the power problem more keenly, been more specific in his recommendations as to its control for the public benefit." He has fought to see that "the domestic and rural and other small consumers shall reap the benefits of cheap power. Governor Roosevelt's announced power program . . . seems to me to include everything that is vital."[37]

Comparison of Giant Power and SuperPower demonstrated how the power industry fought any plan threatening its prerogative in the rural market. When industry, home and nearly every phase of American life were converting to electrical energy, agriculture was, therefore, an exception. The Giant Power Board had tried to correct the situation, and despite its defeat, ideas and observations on the nature of rural service were made. For Cooke the electrification of farms became a special concern, and he was recognized as the most literate person on the subject. His contact with Roosevelt enabled him to carry his ideas to the one man who could possibly get action, for with Roosevelt's victory in 1932, Cooke had access to and was part of the federal policy-making apparatus. Opportunity to put his plans into action had never been better.

4
CREATION OF
REA—1935

For the first two years of Franklin D. Roosevelt's New Deal, Morris Cooke worked as a consultant in the Public Works Administration (PWA), but his real job was to advise the president on conservation and power matters. He never lost sight of the need, however, to provide the country's forty million rural inhabitants with electric service. Shortly before the president's inauguration Cooke recommended that "as part of public works construction we undertake to carry electrification to 50 to 75 percent of our farmers." Such an expansive program, he continued, could not be accomplished in a short time, but "a large part of this could be accomplished in your term in office."[1] Roosevelt agreed that every home and farm should have service, but he showed no willingness to commit himself to any plan that would carry out that objective. Cooke sensed, nonetheless, that if Roosevelt were convinced of the technical and economic feasibility of rural electrification, he would agree to launch a federal program to serve farms. For the next two years Cooke was the administration's chief protagonist for such a program.

Cooke was not the sole instigator in the creation of REA, however, for Roosevelt's establishment of the agency was a reaction to a series of events that occurred in 1935. The successful completion of an experimental cooperative by TVA removed much of the doubt about public electrification. Two states, furthermore, had already outlined programs to serve their farmers and had pressured Roosevelt for federal funds to begin their plans. Politically powerful agricultural organizations, voicing the desire of their constituents for electricity, also brought pressure to bear, so that in 1935 Roosevelt, with Cooke's prodding, created REA by executive order.

When Cooke had made no progress with his personal plea in February to the president, he sent a proposal to the National Recovery Administration (NRA) for a joint public-private program with the electrical industry. Presidential secretary Louis Howe replied that NRA would review his suggestion, but Cooke heard nothing else. In the meantime, however, Howe informed the president that "there appears to be an opportunity to inaugurate partnership arrangements between the public authorities and the private interests by which he [Cooke] can electrify rural America now."[2] Cooke was not optimistic and he told his friend Harry Slattery, personal secretary to Secretary of the Interior Harold Ickes, that when the proposal reached NRA "a laugh will result."[3]

Cooke's proposals did not bear fruit, possibly for several reasons. In his own words, they were "academic unless one knows what he discussed with utility executives."[4] The president had already devoted considerable resources to rural problems via the Agricultural Adjustment Administration (AAA); the same was true for public power with the proposed TVA. Another plan as large and controversial as Cooke's might bring enough unfavorable reaction to jeopardize other programs that were equally pressing. Roosevelt, too, may have anticipated the creation of a separate agency for rural electrification and refused to become involved with private interests, although he never clarified his views. Cooke made one last personal plea in July when he wrote presidential secretary Marvin McIntyre that "we appear to have a plan under which this work can be executed at fast tempo and with the cooperation of the electrical industry if the President is sufficiently interested."[5] He received no response from the White House; the silence was a setback. Thinking the president might respond favorably to pressure if it reflected a sense of organization and unity rather than a personal plea, Cooke took his fight to the cabinet and agency level where he hoped to build support for his idea.

In the meantime, TVA began operating an experimental cooperative that established the all-important first proof of the feasibility of publicly-developed rural electrification, the Alcorn County Electric Cooperative. A more appropriate place to test the effectiveness of cooperatives could not be found, for Alcorn county was located in northeastern Mississippi, a depression-ridden spot with a high number of tenant farmers. Only 1.5 percent of Mississippi farmers had electricity. Such a place well-represented the one-crop South.[6]

Throughout the debates over the development of the Tennessee River, the proponents of public power had not specifically demonstrated how to implement their plans. When TVA went into operation, therefore, no one knew how to solve the Herculean task of getting electricity to farms and rural homes. It was in the rear of a furniture store in Corinth, Mississippi, county seat of Alcorn, that TVA officials, including TVA director David Lilienthal, met with several townspeople and agreed to form a cooperative. Lilienthal implied that this "unpretentious beginning of farm electricity co-ops" was spontaneous, but some preliminary discussions between the parties had already been held. The decision to resort to a cooperative was inspired by the approximately fifty examples already in operation which had given partial credence to the idea that the "self-help" method would work. But it was true, as Harcourt A. Morgan, chairman of the TVA Board, admitted, that these examples had been "tried under exceptionally favorable conditions."[7]

Operations at Alcorn began in June 1934 when the cooperative entered into a contract with TVA to purchase power at wholesale rates. Corinth, a town of 6,220, had been served by Commonwealth and Southern (C&S) which had already ceded the territory to TVA, but farmers in the area had never been served. The cooperative furnished power to the townspeople as well as farmers, an important exception to the expected normal method of operating an electric co-op. Energy was furnished to Corinth residents at a 50 percent reduction of the C&S rate and service was extended to farmers in the environs. The latter did not pay a surcharge as was normally the case when served by a company. Anxious to take advantage of the conveniences suddenly available to them, rural homeowners purchased appliances as quickly as possible, and town merchants reaped a rich harvest in the sale of electrical merchandise.

The real crux of the experiment, the question of solvency, proved to be no problem. Within the first six months of operation, the cooperative had gross revenues of $38,460 and a net income of $14,435 after deducting taxes, interest, depreciation and other expenses. New income amounted to 37 percent of the gross. At such a rate of progress the co-op's indebtedness to TVA could be paid in five and one-fourth years, rather than the projected twelve to fourteen years. Inclusion of the city of Corinth, however, was a special advantage because it permitted a lower rural rate than would have been charged, although TVA officials were reluctant to admit it.[8]

Several accomplishments of the Alcorn experiment demonstrated the feasibility of the cooperative as a means of extending electricity into rural areas. Farmers, as Cooke and others had suspected, used more energy than city residents, which helped offset the extra cost of building rural lines. Area coverage was possible, because the prosperous farms absorbed the loss from those producing low revenue. Co-op users, farm and non-farm, promptly paid their bills. "The fact that the system has already demonstrated financial solvency," wrote one observer, "is evidence of the soundness of this form of organization."[9] Cooke kept a watchful eye on Alcorn and remembered its success when he organized REA. Looking back years later, he wrote: "This early experience with the cooperative device proved of definite value to the development of the present national rural electrification program."[10]

A critical ingredient in the success of the Alcorn Cooperative was the Electrical Home and Farm Authority (EHFA). One of the obstacles to rural electrification was the initial investment and cost to the consumer of acquiring appliances and machinery. To overcome this hurdle, Roosevelt created the EHFA in December 1933 as an emergency agency. It was Lilienthal's "pet creation," an experimental undertaking in cooperation with TVA. Its purpose was to enable residents to acquire appliances and use more electricity.[11]

EHFA made agreements with the manufacturers of electrical appliances in the specifications and standardization of three items—ranges, water heaters and refrigerators. Arrangements were made with power companies and cooperatives to act as retail distributors. Each customer went to the distributor, signed a purchase contract and made the first installment payment. EHFA paid the retailer the balance. The purchaser's unpaid balance, however, was divided into installments and incorporated into the monthly electric bill.[12]

The key to the plan was low-cost financing. A single appliance was financed over a period of three years and two or more appliances over a period of three to five years. Private lending institutions at the time normally placed the maximum lending period for appliance loans at twenty-four months. EHFA interest rates were set at 5 percent, about half that generally in effect for small consumer loans. Congress did not make appropriations for the agency since it obtained funds from regular private sources and from the earnings of the business.[13]

EHFA loans were made available throughout the TVA area of operation; appliance sales jumped 300 percent. Greater consumption of energy enabled TVA to offer minimum rates. By combining low cost energy with loans for appliances, TVA broke the cycle of expense of the rural electric customer and put electrification within the range of the farmer's pocketbook.

Elated over the success of his creation, Lilienthal described the public acceptance of the agency as "one of the most heartening things in American life."[14] So successful was EHFA that it was extended in 1935 to cover the entire United States, although its accomplishments on a nationwide basis were considerably less dramatic. Cooke was appointed to the board of directors in 1935, but he soon resigned because he regarded the board's policy as too conservative. Perhaps he was too hasty, for EHFA proved successful until 1943 when it was dissolved because the government no longer encouraged the sale of appliances.[15]

Opportunity to take advantage of modern technology was the basic reason for the enthusiastic response of the rural inhabitants both to TVA and EHFA. It was no accident, however, that northeastern Mississippi was the first to act, because Alcorn county was in the congressional district of Representative John Elliot Rankin, the most polemical public power enthusiast in Congress. Long known mainly as "an advocate of white supremacy," Rankin was first elected to Congress in 1920 and served until his defeat for renomination in 1952. He co-sponsored the TVA bill with Senator Norris in 1933, later participated in the passage of the Rural Electrification Act in 1936, and was described by REA administrator John Carmody as "probably the best informed man on public power in the House of Representatives."[16] His home town of Tupelo had the first wholly municipal electric cooperative served by TVA, and his district was the first in the United States to have total public electrification, rural and urban. Rankin was convinced that power companies had no social conscience and that the electrification of the countryside was possible only through public enterprise. "The greatest development yet undertaken by the TVA," he would say, "is that of rural electrification. We expect to light every home in that section of the country and to give the farmers electric lights and power."[17]

Success of the Alcorn cooperative gave the proponents of public power a sense of triumph. It demonstrated that with low rates and availability of appliances, the consumption of electricity rose and the cost went down while improving the standard of living. For years such a development had been anticipated, even while the utilities insisted that the farm use of energy was an educational problem with the consumer. With the first federally-sponsored co-op successfully established, the operation of public electric cooperatives on a wide-scale was a step closer to reality.

Meanwhile, Cooke worked to secure President Roosevelt's approval of a nationwide program. He worked simultaneously on several projects for Roosevelt, some of which did not bear directly on rural service, but he took every advantage, nonetheless, to convince administration officials of his plan. Cooke concentrated on Secretary of the Interior Harold Ickes. On several occasions he presented Ickes with a proposal for a joint program with the electrical industry. In one instance Cooke discussed his plan while Ickes listened in silence. The time was ripe, Cooke urged, because the power companies were willing to help. The "Old Curmudgeon," still smarting from a rib broken a month earlier, suddenly told Cooke: "I'll have nothing to do with the sons-of-bitches."[18]

Cooke was not prepared for his reply, but sensing an opportunity, he asked, "Then will you consider a plan wholly under control of public authority?" Ickes responded by asking if public electrification was feasible. Cooke quickly replied in the affirmative, and Ickes, in his usual style, answered. "Then shoot!"[19]

This meeting set off the chain of events within the administration that ultimately led to the creation of REA. Cooke had hoped to make arrangements with the power industry since they already had the resources for the enormous task. Ickes did not, however, share his friend's optimism about help from the power companies, and his blunt answer clearly indicated that a joint venture was out of the question. Ickes may have been overly pessimistic because conditions for cooperation, according to Cooke, were encouraging. An impressive list of power company officers had promised to abide by government ruling on rates and area coverage, and their meetings with Cooke had been warm and friendly. Cooke also said it was power company executives who had "made the suggestion that it might be worthwhile to have a commission set up, half government and half private power with a governmental chairman."[20] But in view of Ickes's blunt answer, Cooke immediately went to work on a public plan.

Cooke organized a series of meetings of his friends and associates at the University Club in Philadelphia to draft a new proposal. In Cooke's words, "many new phases of the rural electrification subject" were taken up.[21] It might be disastrous, he decided, to submit another memorandum on what some people saw as a dry and dreary subject. He was determined, therefore, to prepare a report that would go beyond the ordinary, one that would receive attention. From Leland Olds, still with the New York Power Authority, he obtained the services of an artist and draftsman to assist in preparing the final report. The artist used water colors and three sizes of type "so that," as Cooke stated, "he could take a rather ordinary document and doll it up so that it stood out."[22] They discovered at the last minute that Ickes liked red barns, so every barn shown in the report was a brilliant red. The cover was painted with a black and white zebra-striped design "so flamboyant," Cooke later wrote, "that I remember saying to myself, 'This will keep them from throwing it in the waste basket.' "[23] On February 13, 1934, Cooke submitted the all-important document to Ickes. "To safeguard your time," Cooke wrote in an attached letter, "I have written what amounts to a foreword which can be read in ten to twelve minutes."[24]

The report proposed a Rural Electrification Agency within the Department of the Interior "manned by socially-minded electrical engineers." Self-liquidating cooperatives should be used because their feasibility had already been proven in the assorted examples throughout the United States. Construction of lines would be financed by the proposed agency and technical, engineering and management advice would also be furnished. So as not to compete with the electric companies, Cooke wanted the cooperatives to serve only those areas not already supplied. Since only 10 percent of American farms had service, the plan was massive, representing one of the largest public works proposals within the administration.[25]

Low-cost financing was the key to the plan. Long-term loans to the cooperatives with minimum interest would enable them to build lines without resorting to prohibitive rates. In addition to his regular monthly bill each customer would pay $1.00 to $1.25 to cover his share of the construction cost, a more realistic financial commitment than the $500 deposit then required by the power companies. To promote consumption of electricity, which had to be about 100 kilowatt-hours a month per home in order to effect rate reductions, Cooke urged low-cost financing of home appliances as practiced by EHFA; rates should then fall within 2 to 3 cents per kilowatt-hour.[26]

Cooke did not indict the electrical industry. Recognizing that cost had been the drawback to further development, he wrote: "Large average use, especially in the initial stages, seemingly requires planning and investment beyond the capacity of a private company to initiate. Perhaps only the power and force of the government can master the initial problem."[27] In his earlier proposal for a joint program, Cooke believed the "companies would like to make this jump, but dread the years during which the transition is taking place."[28]

With its unusual format and plainly written explanation of a technical subject, administration officers quickly read the report. S. H. McCrory, chief of the Bureau of Agricultural Engineering, told Rexford G. Tugwell, assistant secretary of agriculture, that "the program outlined in the report is most interesting and . . . I believe feasible."[29] Secretary of Agriculture Henry A. Wallace informed Ickes that "I think the program along the lines sketched out here is well worth considering, particularly in connection with a program of farm rehabilitation."[30] Secretary of Commerce Daniel C. Roper and relief administrator Harry Hopkins also liked the proposal.

Cooke later referred to the report as "the detonating force which started rural electrification."[31] Favorable reaction to the plan was due to the growing realization within the administration that rural inhabitants deserved the benefits of modern living as well as other people. Cooke ably demonstrated how the government could sponsor a sorely needed and complex program. Using language free of technical jargon, he emphasized what running water and indoor bathrooms would mean to rural families, something easily grasped by cabinet officers and bureaucrats.

Administration officials had always regarded electrification as a desirable step toward improving rural life. Secretary Ickes had used PWA funds for extending lines on a few relief projects and for building small diesel generating plants and street lighting systems in a few country towns. In 1933 the Federal Emergency Relief Administration (FERA) had started conducting surveys of areas that could be served, continuing until August 1935 shortly after REA was created. The Civil Works Administration (CWA) Farm Housing Survey demonstrated the widespread lack of modern conveniences in rural homes. These various surveys, which duplicated each other, demonstrated the administration's cognizance of the need for rural service.[32]

Even though such piecemeal action showed general concern within the administration, Cooke received no commitment to carry out his plans. Nor did his zebra-striped report with its popular appeal persuade Ickes or anyone that the administration should undertake such a program. Disappointed, but refusing to quit, Cooke saw another opportunity to promote his idea when the Mississippi Valley Committee (MVC) filed its report on October 1, 1934.

A year earlier Roosevelt had created MVC as a special panel of PWA with Cooke as chairman. The president ordered a full-scale study of the Mississippi Valley, with recommendations for its development so as to improve the quality of life of the inhabitants. So extensive was the study and so great was Cooke's enthusiasm for it, that MVC occupied the major portion of his time for nearly a year.

The report pointedly illustrated how the misuse of natural resources— soil erosion, undeveloped water power, exploitation of timber and poor farming practices—could be responsible for losses in human opportunity. To overcome the waste of humanity on the richest farmland in the country, MVC urged federally coordinated scientific management of resources in the valley.

The MVC report bore the imprint of Cooke's influence; it focused much attention on rural electrification. Only 10 percent of the farm families had service in an area blessed with abundant water power; yet the report regarded electricity as the most important element in the rehabilitation of rural life. It would advance the agricultural strength of the nation, and it promised enhanced social well-being and home comfort for thousands. Specifically, the report recommended federally financed and supervised cooperatives. To begin work Congress would have to appropriate $100,000,000. Power companies were excluded from the program. In his "zebra-report," Cooke had called for the use of cooperatives, but did not specify the amount of money necessary to begin construction.[33]

Rural electric development was only part of the larger MVC composite, for it urged multi-purpose planning in the valley. So well-prepared and impressively written was the report that it caused considerable stir. *The New York Times,* to quote one writer, "was impressed by the vision in the Mississippi report. . . ."[34] For a short time MVC gave a lift to the concept of multi-purpose planning and Cooke hoped to get action. None came, however, and he was again disappointed.

Unwilling to let his project be tabled, Cooke continued his campaign to persuade officials. As chairman of the Water Planning Committee of the National Resources Board which superseded MVC, Cooke reported that "industries have recognized the use of . . . electrical power. Agriculture has lagged . . . it therefore seems necessary for the Government to stimulate the extension of this service in many areas."[35] He told the president and Ickes that the National Power Policy Committee, of which Cooke was chairman, needed guidance on the subject. His repeated urgings, tirelessly saying the same thing, had one purpose—namely, to prod the president into action.[36]

In December 1934, Cooke took a month's vacation in Phoenix, tired from the MVC assignment and the assorted committee work assigned to him. He had bombarded the administration with plans for either a public or private program or a combination of the two and had progressed in the sense that the electrification of rural America was recognized as a fundamental problem crying for a solution. As far as seeing the practical application of his proposals, however, he went to Arizona empty-handed. The breakthrough via the Alcorn co-op was just becoming clear when Cooke left, and he could not be sure that the steps made at Alcorn would be expanded throughout the nation. For Cooke the question was not whether electrification was technically or economically feasible but whether the president wanted to commit himself to such a program. It seemed that Cooke's proposals were ignored by the occupant of the Oval Office.

Although Roosevelt recognized the importance of electrification in improving rural life, he had never given Cooke any hint of direction. Whatever Roosevelt's thoughts, he found himself increasingly under pressure to act. In 1934 the American Farm Bureau Federation and the National Grange had passed resolutions at their annual meetings urging federal action. The former resolved that the Farm Credit Administration should finance cooperatives and mutual light and power associations. The Grange wanted a system that would "deliver power to the people . . . at the lowest cost possible."[37] Farm Bureau president Edward O'Neal had reminded Secretary of Agriculture Wallace in September 1934 "that this rural electrification problem is, in many respects, a national problem and that we should have some help."[38] A short time later the president invited O'Neal to the White House for a private conference on power matters relating to agriculture. O'Neal told him: "Appoint this big boy from Philadelphia to advise you about rural electrification. That will revolutionize the standard of living of rural people like nothing else could."[39]

Pressure on the president came from still other sources. North Carolina and South Carolina had arranged to serve their rural citizens through state operated systems similar to that in Ontario. Cooperatives would not be used; farmers would be supplied directly by state owned and operated lines, and power companies would furnish the energy. So far advanced were the sister states in their preparations that each had an application for federal funds to begin construction pending before PWA. South Carolina had been waiting since December 1933.[40]

By late 1934, PWA had not acted on either application, and the two states were angry over the agency's reticence. A South Carolina delegation went before PWA in hopes of expediting its case, and in December 1934, North Carolina Governor J.C.B. Ehringhaus visited Roosevelt at the White House, also hoping to expedite approval of his state's application. In neither case was the personal visit successful, but the pressure was significant in convincing Roosevelt to take action on the subject of electrification.[41]

At the beginning of 1935, conditions were, therefore, ripe for Roosevelt to commit himself. In his annual message to Congress in January, the president recommended a program as a relief project to make possible "new forms of employment." In April when Congress authorized nearly five billion dollars for public works, it allocated $100,000,000 for construction of rural distribution lines. Congress settled on that amount because it was the figure suggested by MVC for starting a program in the Mississippi Valley.[42]

The president decided to establish a separate agency to disburse the $100,000,000, but he had not decided who should direct the program. At a press conference April 24, he told reporters, "I have not talked to anybody about it; I have nobody in mind yet."[43] He had, however, instructed Frank R. McNinch, chairman of FPC, to check on Stephen DuBrul of General Motors as a candidate. But in his report, McNinch suggested that the president consider Cooke who planned to sail on July 1 for three months in Europe. Roosevelt instructed his secretary, Stephen T. Early, to arrange a meeting with Cooke. The chief executive persuaded Cooke, probably with little difficulty, to accept the new post because a few days later he told Roosevelt "that the new Rural Electrification Unit is a going concern. A suggested draft of the executive order officially setting up the REA will be in your hands early in the week."[44] The order arrived as promised, and Roosevelt officially created REA on May 11, 1935.

A convergence of events at the end of 1934 was responsible for the commitment to electrify farms and rural homes. In keeping the matter before administration officials, and in resolving certain technical and organizational problems, Cooke was the driving force. Success with the Alcorn Cooperative, which came to light only by late 1934, reinforced his argument that service was feasible on a national scale. It was evident, furthermore, that some states were ready to move and preempt the federal government if Roosevelt did not act. Finally, the stance taken by the Farm Bureau and Grange reflected the growing militance of the rural population for assistance in their efforts to modernize home life.

If the drive to furnish every farm and rural home with electricity had reached a new plateau in 1935, practical use of the current was still in the future. It remained for Cooke to implement the policy that was now his responsibility. He proceeded to investigate every alternative for serving farms.

5

THE SEARCH FOR AN OPERATIONS PLAN: TRIUMPH OF THE COOPERATIVE

During its first year of operations, May 1935 to May 1936, REA was a temporary agency and furnished only a handful of farms with service due to the lack of an operations system. Morris Cooke tried to arrange a joint program with the power companies because he knew they were already equipped to build distribution lines, but they would not agree to follow his conditions. At the same time, the power industry clashed with the Roosevelt administration over the "death clause" of the Utility Holding Company measure of 1935 which killed any chances for cooperation between REA and utility companies. To a lesser extent Cooke tried to work with municipal power districts and again was rejected. He reluctantly turned to cooperatives, selecting them by default rather than choice. Cooke's difficulty in developing an operations plan, even after the federal commitment was made, demonstrated the extent of the administrative barriers REA had to overcome.

Cooke first had to establish a procedure to carry out the directive that "in so far as practicable, the persons employed under the authority of this Executive Order shall be selected from those receiving relief."[1] Since other agencies involved in the electrification of farms prior to 1935 were relief agencies, it was assumed that REA would proceed essentially along the same lines. As a relief agency REA also had to spend one-fourth of its funds directly on labor, of which all but 10 percent had to come from unemployment rolls. Cooke immediately discovered that the policy regarding labor was impossible because skilled workmen were required in the construction of electric lines and the manpower pool on relief rolls was unskilled. Also a substantial portion of REA funds had to be spent on equipment and material. "A very small percentage of our outlays," Cooke admitted, "went for unskilled labor, the class then in direct need."[2]

Struggling to find some way to carry out the relief provisions of the executive order, Cooke met with Roosevelt, Ickes and Harry Hopkins. They agreed that REA could not proceed as long as it was saddled with the responsibility of furnishing relief to the unemployed, and arrangements were made for it to proceed as a lending agency. The comptroller general approved the legality of the proposed change, and Roosevelt issued an order giving the agency the exclusive right to make loans to power companies, public power districts and local organizations engaged in rural electric construction. Freed from the restraints concerning relief spending, REA could proceed with its primary work. H. S. Person, economist with the agency, later wrote that "this was probably the first and most far reaching fundamental policy decision in the history of the REA."[3]

What proved to be more perplexing than provisions concerning relief, however, was the search for a plan of operation. Cooke faced three possibilities: REA could loan money directly to the power companies and let them build and serve the lines; it could lend money to the municipal electric districts which could do the same; or lend directly to specially organized cooperatives. Neither the president nor Congress in the Emergency Relief Act of 1935 that funded the new agency had given any direction on the matter. The National Power Policy Committee did not indicate any preference.

Sentiment for the use of cooperatives was particularly strong in public power circles. Judson King published a bulletin entitled, "Who Will Get the $100,000,000 for Farm Electrification?" urging farmers to organize cooperatives and asserting that if they went through the utilities they would pay "another series of mortgages . . . in the shape of watered stocks for operating and holding companies. . . ."[4] Senator Homer T. Bone of Washington insisted that if the power companies received the appropriation, the REA program would "not be of very much value to the farmer."[5] Congressman Rankin did not want the funds to fall into the hands of the "silk-hat brigade of Power Trust manipulators and watered-stock brokers."[6]

Momentum was, however, in favor of the private interests. Cooke had to spend the $100,000,000 quickly and only the companies had the equipment and personnel to go into action on short notice. Interest in the subject was still strong in utility circles. As late as January 1935, a committee of the industry's leaders had submitted a one-year program to Cooke when he was still a PWA consultant. Utility executives knew that Cooke was not a rabid opponent of the industry, for though he agreed it had not fulfilled its obligations to the rural citizenry, he did not agree with those who wanted government ownership of the electrical industry. Rather, Cooke belonged to the school of regulation, wanting strong and honest state public service commissions, legislation to control or restrict holding companies and public officials able to withstand pressure from special interests. In short, Cooke saw the industry as a potential ally and not the enemy.[7]

It was this reason that prompted Cooke to invite representatives of the electrical industry to meet with him at the LaFayette Hotel in Washington on May 20, 1935. According to Person, Cooke hoped "the companies would be moved by a sense of public duty," and would have a desire for participation "strengthened by the low interest rate of three percent and other generous terms offered by the REA."[8] The LaFayette meeting was friendly: utility representatives agreed to appoint a special committee to survey "the approximate extent to which further development of rural electrification may be promptly effected in cooperation with the REA."[9]

After the meeting it was anticipated that a joint plan would soon be drawn up. Several editorials in *Electrical World,* trade magazine of the power industry, pointed to a spirit of friendliness and cooperation among utility executives. Grover C. Neff, president of Wisconsin Power and Light, advised colleagues in the industry to cooperate and show a "pioneering spirit." *Business Week* reported that Cooke was "entirely agreeable to having private capital assume as much of the burden as it wants."[10]

But each side had reservations which had either been overlooked or misunderstood. On several occasions when prominent utility executive Hudson W. Reed, of the United Gas Improvement Company, stressed the need for cooperation with REA, he maintained a sense of caution and reserve. At a conference of the Edison Electric Institute less than a month earlier (previously the NELA), he warned industry leaders "not to jeopardize your potential rural business by permitting governmental yardsticks to be established in every section of the country."[11] Further, he told *Electrical World* that while the industry "will make every possible effort and sacrifice to further this desirable social program . . . the industry should know how far it can proceed consistently with sound business practice."[12] In an interview with the same journal, Cooke emphatically asserted that power companies would have to follow a self-sacrificing policy if they expected to receive REA money. Only if the government assumed active leadership, he repeated for *Business Week,* could a significant portion of farms receive service. In one sense a friendly and cooperative spirit was evident, but prima facie evidence also indicated some strong reservations on each side, perhaps enough to result in a direct clash.[13]

The showdown came when the specially created committee of the industry met with Cooke on July 24, 1935. The utility executives submitted a plan proposing REA loans to power companies totaling $113,685,000 for 1935-36; it would connect 351,000 prospective rural customers of which 247,000 were farmers. The cost per mile averaged $1,356. Electrification according to the executives was "a social rather than an economical problem." In regard to rates they reported that "the problem of the farmer is not one of rates but of financing the wiring and purchasing of appliances." Farmers, they continued, were the "most favored" class of customers and "as a result, there are very few farms requiring electricity for major farm operations that are not now served."[14] The committee did not discuss rates and area coverage, the two critical barriers to development.

Cooke and critics of the industry were gravely disappointed. How could it be asserted, they asked, that only a few remaining farms had no electricity when only 10 percent had service? How could it be argued, they further asked, that rural residents received preferential treatment when they paid the highest rates? Nor could critics agree that farmers should be blamed for their lack of service because they could not buy appliances or wire their homes. Critics regarded the report as another example of the industry's refusal to accept responsibility for developing the rural market. The chief weakness, wrote Person, was "the lack of appreciation of the significance of rates and other costs to progress in rural electrification."[15]

Cooke's reply was short and noncommittal. He indicated his desire to cooperate, but wrote: "we hold rate simplification and even rate reduction over large areas to be the heart of the problem of electrifying rural America." On the question of financing the wiring of homes and purchase of appliances, he suggested the committee consult EHFA. Cooke invited companies to submit applications for loans on an individual basis, but made no commitments to grant loans.[16]

At the heart of their divergent views was the definition of profitable service. Utility executives used an interpretation of the rural use of electricity different from Cooke's. For them it meant the use of power in large operations such as dairying, irrigation or poultry farming in which electricity as a unit cost of overhead was less expensive than alternate sources of power, namely animal or hand labor. Within the power industry, the household use of electricity was not measured in terms of profit and investment, and although utility executives acknowledged electricity's social and personal advantages, they did not regard it monetarily cheaper than traditional hand labor. The executives also failed to realize that as families acquired more appliances their consumption increased, making lower rates feasible, a lesson already proved by the Alcorn Cooperative. Since electric companies insisted on recovering construction costs in a short time, they charged such high rates that only farms with large-scale operations could use electricity profitably, and indeed the industry had already extended service to them. Holding this view of profitability, the committee was correct in reporting that "few farms requiring electricity for major farm operations are not now served." Cooke was disappointed in the committee's narrow definition of the rural market and thought it reflected an old-fashioned view, limiting the use of profitable electrical energy to manufacturing.

The July meeting proved to be the last chance for a joint program. Any possibility for rapprochement was definitely lost in the battle over the Public Utility Holding Company Act of 1935. Known as the Wheeler-Rayburn bill for its sponsors, Senator Burton K. Wheeler of Montana and Representative Sam Rayburn of Texas, it was introduced into Congress on February 6, 1935, and ignited a bitter legislative fight. The administration had initiated the measure believing that regulation of the holding-company would correct the numerous "ills" of the power industry. The bill did not bear directly on rural electrification except in the sense that regulation and elimination of profit-draining holding-companies would allegedly encourage development of the rural market.[17]

The measure was responsible for an unusually tough battle between the administration and the power trust. To kill the measure, wrote one historian, "the private utilities subjected the members of Congress and the public to one of the most carefully planned and executed lobbying campaigns in American history."[18] Congressmen received thousands of telegrams denouncing the proposal, sometimes signed with names taken from telephone directories. So numerous and acrimonious were the charges fired back and forth during the congressional debate that the Senate and House each established a special fact-finding committee to investigate all sources of influence on the bill. The Black Committee, named for its chairman Senator Hugo L. Black of Alabama, uncovered "an incredible story of lobbying, pressure politics and intrigue."[19] The power industry saw itself fighting for its life due to the "death clause" in the proposed legislation which would eliminate any holding-company that could not justify its existence. So emotional was the atmosphere that some party regulars deserted Roosevelt and forced him to accept a compromise. The "death clause" was amended enough to take the sting out of the bill, even though trust-busters regarded it as a triumph.[20]

Cooke was trying to work out his agreement with the electrical industry when the furor over the measure occurred. Success was practically impossible. At one point, Hudson Reed stated that "since the introduction of the holding company bill, negotiations [with Cooke] have lagged;" another utility executive warned his colleagues "to proceed cautiously" with REA. Opportunity for cooperation, Cooke's hope since his days with the Giant Power Board, was gone.[21]

Throughout his negotiations with private interests, Cooke sought other ways to establish an operations plan. Since some farms had received service from municipal power plants, Cooke saw them as a possible instrument. On May 24, 1935, four days after his first meeting with the utility executives, he had a preliminary conference with representatives of municipal power systems. It was agreed to hold a joint meeting of municipal and REA spokemen in Kansas City, Missouri, on November 7-8, 1935.[22]

One hundred fifty-two delegates from seventeen states attended the meeting. W. E. Herring, Cooke's special assistant, explained the possibility of serving farms through municipal electric districts. The meeting was useless. Representatives of the cities were not concerned about the needs of the farmers; some anticipated higher rates for urban residents if service was to be extended into the environs, while others worried about the question of jurisdiction between cities over intervening territory with potential rural sales. Possible legal entanglements between cities and their respective state legislatures over rural service also caused some anxiety. The solution for REA lay elsewhere. "I did think that we were going to be able to do business with the municipalities," Cooke told his staff, "but they turned out to be more difficult to do business with than the private companies."[23]

After six months of searching, Cooke had found no way to carry out REA's responsibility. Whereas originally he had hoped to work with the electric companies, he now anticipated lack of cooperation if not opposition. Efforts to work through municipalities led nowhere. By autumn of

1935, therefore, the agency was no closer to an established plan of operation than when it opened its doors in the spring. Only cooperatives were left.

Opinion varied over the cooperatives. Some REA staff members opposed them; the power contingent in Congress, plus other REA staff members preferred them. Cooke saw use for cooperatives only on a small scale, applicable in those instances wherein a group of farmers were especially prosperous and showed ability to handle the legal and financial affairs. The average farmer, he thought, would require more assistance than REA could provide; power companies with REA loans and supervision were best suited to assume the task.

Cooperatives had a history and tradition, however, that were advantageous. American farmers had long experience with them in the area of marketing, and those electric co-ops in operation as far back as 1913 gave credence to their use. The Alcorn Cooperative had suggested that with proper guidance farmers were capable of serving themselves.

Public power proponents were keenly aware of these successful experiments and had applied pressure on REA for adoption of the cooperative. Representative Walter M. Pierce of Oregon told Cooke: "People are forming utility districts to take advantage of opportunities suggested by the establishment of your administration. Their confidence would be increased if they were offered substantial evidence of government interest in their enterprise."[24] Rankin was even more confident of cooperatives once he was armed with the facts and figures of the Alcorn Co-op in his own district. At the Public Ownership Conference in Chicago he accused power companies of browbeating Congress, hiring lawyers for political influence, taking bonuses and "rake-offs" and buying up space in newspapers and magazines. For Cooke he had a message: "I commend the county unit system because . . . it is the best plan that has yet been developed to reach all the people . . . with cheap electric lights and power. . . ."[25] Not to be overlooked was the pressure applied directly by agricultural organizations. Earlier Murray Lincoln of the Ohio Farm Bureau had asked Cooke to establish cooperatives in his state because he had no confidence in the utilities. Cooke reluctantly agreed to help him, and Lincoln quickly organized two of them. "Farmers were just itching to have electricity," Lincoln recalled, "and to have it from their own cooperative was a dream come true."[26]

Additional pressure for adoption of the cooperative came from the REA Development Section headed by Boyd Fisher, a one-time business partner of Cooke. The Fisher contingent, unlike many REA staff members, believed farmers were fully capable of operating electric cooperatives as long as REA furnished supervision and funding. Fisher steadily prodded Cooke, telling him that municipalities and power companies had other interests, that only the farmers with their overwhelming need for service had the enthusiasm and desire to overcome the obstacles to rural service. But Fisher carefully insisted that REA not engage in the propagandizing of co-ops, but act as a credit agency granting loans only to those cooperatives founded in a solid, business-like manner. The real impetus in other words should be generated locally, not in Washington. Fisher's presence had the effect of keeping the concept of the cooperative close at hand, and with the failure to arrange a plan with the utilities or municipalities, his idea took a step forward.[27]

It was thus a natural outgrowth of the long experience with the co-operative that led Cooke and REA officials to meet with spokesmen and representatives of agricultural marketing cooperatives on June 6, 1935, in the midst of his negotiations with the electrical industry. Delegates expressed concern about the feasibility and practicality of the proposed co-op since it would require the skill of engineers and legal personnel, and a host of services farmers were unable to provide for themselves. It was explained that REA would furnish these services. Joseph C. Swidler of the TVA legal staff recounted the story of the Alcorn Cooperative, explaining how it proved to be a "stable and efficient operation unit." He continued: "Besides it is the farmer's only answer to the failure of public or private plants to bring the benefit of cheap electricity to him."[28] With the precedent of the successful Alcorn experiment before them, the group appeared willing to try electric cooperatives, and Cooke described the conference as "most encouraging." But no concrete action or decision resulted from this meeting.

A few days after the conference, Cooke asked C. W. Warburton, director of the Extension Service, for the assistance of county agents in stimulating and promoting interest for REA services at the local level. "County agents," he said, "are in a particularly good position to confer with and advise farmers and other rural people. . . ."[29] He also encouraged Fisher, saying, "It is desirable to winnow out those local situations where we can depend upon aggressive leadership."[30] Fisher was sent into the field in midsummer to promote cooperatives even as the committee of utility executives prepared their report, and Cooke publicly urged farmers to write for information about establishing co-ops. All of these events spurred demand for action, causing farmers to think that if they went to REA for help, they would receive it.

Cooke gave no signal, however, to begin operations with cooperatives. Fisher received no further instructions, and farmers, despite their pleas, were given no reason to anticipate approval of cooperatives should they act to establish them. From mid-July to November, REA drifted. Cooke persistently waited for power companies to submit applications for REA loans because he still hoped to arrange a partnership with them.[31]

The turning point came when Cooke received an application from Wisconsin Power and Light. He rejected it. Rates of the proposed project were too high and no provisions were made for area coverage. The fact that Wisconsin Power and Light was involved had special meaning because Grover Neff, president of the corporation, had taken a greater part in rural electric affairs than anyone in the industry. He had served on the original NELA committee that recommended the creation of CREA and had been one of the more active participants of the latter. He also was a member of the committee of utility executives that met with Cooke in July. Neff had an understanding of the subject few men could match. For Cooke it was particularly disappointing for Neff to submit an application taking no cognizance of rates and area coverage. Rejection of the application reduced the chances for more requests. Neff warned his friends to thwart REA; it wanted only "to reduce existing farm rates," he insisted, "for political purposes."[32]

Cooke now had to rely on cooperatives. Early in November 1935, he approved loans to eleven proposed cooperatives. The loan interest was set at 3 percent, and the loan had a twenty-year amortization. EHFA arranged to make loans to the participants for wiring their homes and purchasing appliances. Energy was purchased from either a power company, municipal plant or the Reclamation Service.[33]

Indication of the new attitude on Cooke's part was seen in a public challenge he made to the electrical industry to demonstrate its willingness to serve rural residents. In December 1935, he published "An Open Letter to the Operating Electrical Industry" in *Rural Electrification News,* the monthly journal of the agency. "It is a propitious time for you to tackle this problem, basic to yourselves, the farmer and the Nation." He suggested the power companies set up proving grounds in which they served the inhabitants and "unleash in these areas the creative genius and organizing power of your technical force." Electric companies, he added, could "make use freely of the several agencies of the Federal Government which are equipped to assist you. . . ."[34] The message was clear: if utilities electrified an area using REA plans and funds, they could test the agency's willingness to cooperate. There was no response, and "it became clearer and clearer even to Mr. Cooke," John Carmody, the second REA administrator recalled, "that the power companies would not do what they ought to do."[35]

With the cooperative's surge to the forefront, the road appeared open for development, but Cooke encountered another roadblock. He had exhausted his funds to borrowers either by commitment or tentative obligation. On September 12, 1935, Cooke along with cabinet members and other agency administrators had met with the president at his New York estate, a meeting referred to as the "Hyde Park Conference." It was here that Roosevelt worked out the budget for unemployment relief for the year. He gave REA $10,000,000, one-tenth of the congressional allotment. Cooke agreed since it was acknowledged that until REA policy was better defined, the dispersal of funds should proceed slowly. Thus, Cooke was short-handed at the end of the year, and in January 1936, he asked Roosevelt for another $10,000,000 "to meet embarrassing demands." Only half the request was granted, and the total expenditures of REA as a temporary agency only came to $14,003,635. FERA administrator Harry Hopkins used the balance for relief purposes.[36]

Since it went into operation nine months earlier, REA had electrified few farms. Although he had hoped to dash rapidly ahead with development, Cooke had proceeded at a snail's pace. The need for speedy movement had motivated his negotiations with the electrical industry since they could move quickly into the field. But failure to arrange a joint program was due not only to disagreements over rates and coverage, but also the fierce struggle over the Public Utility Company Act. Toward the end of the year the pace quickened with the triumph of the cooperative, only to be thwarted again by the use of REA funds for relief by other agencies.

Because of REA's slow progress, dissatisfaction with the agency grew. Cooke knew he had to improve the rate of servicing farms. By October 1935, he had decided upon a new course—to make REA a permanent agency. That would mean statutory authority and regular funding independent of executive trimming or reallotment. Cooke told the National

Emergency Council that "it is difficult to believe that such a program should not constitute a permanent part of the government's activities."[37] When Senator Norris moved to put the agency on a permanent footing, Cooke knew his idea would receive serious consideration.

In a letter to Cooke, dated October 24, 1935, Norris expressed his belief that it was "increasingly obvious that the time is at hand when, as a Nation, we should adopt a more positive program for electrifying the largest possible number of our farms." He wanted REA to be a twin component of TVA, but that the former should extend throughout the United States. "There should be established in this country a general system for the electrification of rural communities which would take into consideration the supplying of electricity to as many homes as possible. What would it take," he asked Cooke, "so that a much larger percentage of rural homes may be electrified?"[38]

Cooke quickly replied. First, he released a statement to the press about the letter, using it as an opportunity to suggest and agree with the senator that "comprehensive rural electrification must come." Next he carefully drafted a formal reply to Norris, but only after the president had seen it and said, "I think this is very good."[39] The present rate of electrification, Cooke told Norris, meant that "three-fourths of our farm people would remain condemned to drudgery" for a long time. With the program Norris proposed, Cooke estimated that 50 percent of all rural homes . . . could be electrified in 10 years at a total investment, private and public, of $1,500,000,000."[40]

This exchange of letters marked the beginning of a new phase in the development of REA. From that point forward, Cooke and his staff anticipated a smoother operating and more efficient REA if Congress gave it permanent status. Cooke and Norris probably staged their exchange of letters because they had worked closely on power affairs for nearly a generation, and it would have been unusual for one to plunge into something as important as a permanent REA without the other having foreknowledge of it. Public stances such as this one normally came only after Norris, Cooke and usually Judson King had worked out an appropriate strategy. Most likely Cooke approached the senator with the idea in late September or early October. After a half year of frustration and disappointment, REA supporters looked forward to 1936 with renewed hope and vigor.

6
REA
MADE
PERMANENT—1936

The letters exchanged between George Norris and Morris Cooke received front page coverage and gave renewed impetus to farm electrification. At its 1935 convention, the National Grange resolved that REA do everything possible to further service and that funds "be used to accommodate publicly owned power districts, cooperatives and mutual associations in so far as possible."[1] In a similar vein, the Farm Bureau commended TVA and REA for "the use of electricity on our farms, and we urge that this work be expedited in every possible way."[2] No one had more hope and interest than the country folk themselves. A farmer wrote Norris: "I congratulate you on this movement . . . it is the only way that the younger generation will be interested in establishing houses in the rural communities."[3]

Norris called for the establishment of REA as a regular agency with appropriations earmarked for ten years. He wanted a total appropriation of $1,000,000,000, or $100,000,000 per year. At interest rates of 3 percent or less, loans would be made to cooperatives on a self-liquidating basis, repayable within forty years. Only public bodies, cooperatives or municipalities were eligible for REA money, a restriction reflecting Norris's view that whenever possible private enterprise should be excluded from power development. Norris's plan also provided for REA to provide loans to farmers for wiring their homes and purchasing equipment and appliances. Experience with EHFA was responsible for the latter provision.[4]

Norris introduced the REA bill, S.3483, on January 6, 1936 in the Seventy-Fourth Congress, Second Session. It went to the Agricultural and Forestry Committee, headed by Ellison Durant ("Cotton Ed") Smith of South Carolina. He held no hearing mainly because of expressions of support from the Farm Bureau and National Grange as well as the fact that no utility company asked for an opportunity to speak on the bill. As one observer noted, the companies were silent "because of their stinging defeat the year before when the Utility Holding Company Act was passed."[5] At Smith's invitation Secretary of Agriculture Wallace and Secretary of the Interior Ickes commented on the measure, and each urged its passage.[6]

Although the power industry was relatively quiet, it opposed the measure. *Electrical World* described the proposal as "grandiose," and added that the slow record of REA amply proved the futility of public rural electrification. Hudson Reed of the committee of utility executives, who had negotiated with Cooke the previous July, submitted a written statement to the Senate committee. He thought the proposed REA would be prohibitively expensive, that cooperatives would never succeed on a self-liquidating basis, and that the forty-year amortization period on co-op loans was too long. The distribution lines would physically deteriorate in less time and require refinancing of the original loan. The United States Chamber of Commerce wanted the proposal defeated allegedly in order to economize the federal budget, but more likely in response to the prompting of utility lobbyist P. H. Gosden.[7] These reservations on the part of private enterprise had little effect, however, and Smith's committee reported out the measure favorably. Norris anticipated general endorsement from his peers; and Missouri Senator Harry Truman summarized their attitude, saying, "I am very much interested in seeing that the farmers have as many of the good things in life as other people have."[8]

Norris encountered only one obstacle. Congressional leaders and the president considered the proposed appropriation of $1,000,000,000 too high, and the press also expressed dissatisfaction with the figure. Norris admitted the figure was steep and that he had reached the sum arbitrarily. He welcomed suggestions. Cooke recommended $500,000,000, devoting $100,000,000 for appliance and house-wiring loans. Norris met with the president, Cooke, Daniel Bell, director of the Budget Bureau, and Jesse Jones, head of the Reconstruction Finance Corporation (RFC). They amicably agreed on a lower figure of $420,000,000, with the agency getting $50,000,000 the first two years and $40,000,000 per annum for the next eight years.[9]

Opposition to the bill in the Senate was mild. At Reed's suggestion, Norris let the amortization period for co-op loans be shortened to twenty-five years because REA had recommended the same thing. Senate approval came on March 5, and the measure was shuttled over to the House Committee on Interstate and Foreign Commerce, chaired by Sam Rayburn. Except for the reduced appropriation, the bill remained virtually intact.[10]

Rayburn had introduced a similar bill on the same day as Norris, but since the Senate moved faster, the House considered the Nebraskan's measure. Rayburn had been recruited to guide the measure through the House for particular reasons. The lower chamber was generally regarded as conservative, and House opponents of public power had just flexed their muscles by defeating the "death clause" of the Holding Company Act. For the REA measure, a quiet but effective worker was needed, one able to maintain outward calm but likely to succeed. Rankin, the outstanding proponent of rural electrification in the lower chamber, was bypassed because he was considered too radical and might endanger the bill.[11]

Rayburn had experienced the hardships and drudgery of rural life and knew as well as Norris that electrification meant a new way of life on the farm. His family had moved from Tennessee to northeastern Texas in 1887 and bought a small cotton farm. The eighth of eleven children, Rayburn's early life was fully in accord with the rural South stereotype, penniless and filled with self-denial. He first went to Washington to represent the Texas fourth district in 1913, a post he held until his death in 1961. He was best known as Speaker of the House and as an influential figure in the Democratic party. He took the Speaker's chair for the first time in 1940. In 1936 his power and prestige did not equal that of Norris, but he was chairman of the Interstate and Foreign Commerce Committee and became House majority leader the following year. Unlike Rankin, Rayburn had not promoted rural service, but his persuasiveness and quiet manner had already given him a reputation as a particularly effective parliamentarian. In the House, Rayburn was the perfect choice to complement Senator Norris.[12]

Rayburn, however, drew a finer line than Norris in the dispute over public and private power. Whereas Norris fought for supremacy of one over the other, Rayburn wanted to achieve a balance between the two.

He was not a handmaiden of the utilities; he had been the House's chief proponent of the Utility Holding Company Act. He agreed that electric companies were not developing the rural market but saw no justification or advantage in chastising them. His stance on the issue was one of moderation: "We have two schools of thought on this question. One group thinks there should be no public power. I do not subscribe to either. I think that there is a field for both of them."[13] However much Rayburn differed with the senator, other members of the House committee were far more hostile to public power, placing the Texan in the role of conciliator. REA strategists had anticipated such a development, explaining why they had recruited him in the first place.

Committee hearings were held on March 12-14 and involved several spirited sessions with REA administrator Cooke. The committee needled him over the proposed rate of interest, which the Senate had set at 3 percent, and asked him to justify the total exclusion of the electrical industry from the REA program. Some committee members expected interest rates to climb and felt that cooperatives should pay the regular market price for money. Cooke used the same justification as Norris, that 3 percent was the "going-rate" and some industries were borrowing at less. If rates went beyond the REA maximum, the advantage of lower rates would mean money in the hands of the farmers, one of the neediest classes. Such reasoning did not satisfy the committee, and it placed a 3 percent minimum on REA loans, but no ceiling.[14]

Cooke also tried to persuade the committee to accept the Senate provision excluding electric companies. Rural sections were less remunerative, he told them, and "the work of a cooperative can be carried out at much less cost than it could under normally private auspices." The time was getting "nearer and nearer," he continued, "when further electrification cannot be carried on without subsidy." Leave the countryside to the cooperatives, he urged, for they were complimentary parts of the power industry. It was obvious that Cooke had completely abandoned his hope of receiving cooperation from private enterprise.[15]

But the committee saw no reason why the private interests should be excluded. In its view there would be some farmers unable to organize a solvent co-op for various reasons and only a utility could serve them. The hearing ended with a brief plea from Rankin to leave the bill untouched, but the committee amended it to include the utilities and to make the interest rate more flexible. Even then it was approved by a margin of

only one vote. As was the case with the Senate, a committee of utility
executives presented a statement requesting that electric companies be
eligible for REA funds. Only because the industry had just ended the
fight over the holding company bill and preferred, therefore, not to
arouse further hostility, did it not oppose the REA measure vigorously.[16]

Southern sentiment for the REA measure was particularly strong as
seen in the Southern Policy Association's announcement that rural electrification was a major stepping stone toward a better life in the South. The
association was established in 1936 as a Washington-centered group of
southern congressmen backed by "experts" having first-hand knowledge
of the socioeconomic problems of the South. Their goal was "economic
justice" for their region, and their operations were based on the work and
study of southern affairs experts such as Howard Odum, Rupert B. Vance
and T. J. Woofter. A nongovernmental body, the association gave its
members who held public office a chance to mingle with private citizens
having similar thoughts and ideas. It usually held weekly luncheons at
Halls Restaurant at the foot of Capitol Hill, where prominent New Dealers
or practical-minded scholars delivered addresses dealing with southern
problems. Guests and speakers included Harold Ickes, Henry Wallace,
Rexford Tugwell and Morris L. Cooke.[17]

On March 18, only four days after Rayburn's committee had finished
its hearings, the association endorsed a triple package program of legislation to combat the poor quality of southern life. It included the proposed
Bankhead Farm Tenant bill to help tenants and sharecroppers buy farms,
a bill promoting the conservation of natural resources of the South and
the rural electrification bill. Other than announce their position, the group
did nothing to ensure passage of the REA measure. Enactment of the bill
would likely have been accomplished without their fanfare, but the association decided to give electrification national publicity and top priority
for the recovery of the South.[18]

The REA measure touched off a vigorous debate when it reached the
House floor, even though the committee had made changes to the liking
of conservatives. Rayburn was satisfied with the bill and had to defend
it from both sides; conservatives wanted to eliminate it and public power
zealots wanted to expand it. Representatives Schuyler Merritt of Connecticut and James Wadsworth of New York, both members of Rayburn's
committee, fought the measure.

Opponents claimed REA was only another agency in a government
already laden with bureaucracy, and the proposed expansion was another

step toward a socialist state. One man, the REA administrator, would control more than $400,000,000, representing a concentration of power threatening the principle of decentralized authority. Electric companies were afraid to build rural lines because of potential REA intervention, and if this proposal became law, it meant ruin for them. Farmers, so the argument ran, were not experienced in legal and technical matters, and their poor management would cause the cooperatives to fail. In the end the government would be left with a multi-million dollar "dead horse." Better to leave electrification to the private interests than run the risk of incurring a financial boondoggle.[19]

To his opponents Rayburn answered that since electric companies were not serving farmers, the government had an obligation to help them. Solvency of the co-ops would be no problem, he asserted, because farm families so desperately wanted electricity that they would give their whole-hearted support and assistance to them. Rayburn added that RFC had not become a socialist agency, even though it handled large sums of money. The intent of REA, furthermore, was to invest power in local cooperatives, not a Washington bureau. Finally, he reminded them, the bill permitted power companies to participate.[20]

The crux of the battle in the lower chamber was over the electric companies—should they be included? The House committee had said yes, but among liberals a strong sense of moral indignation was evident over the way power companies had treated farmers in the past. Some House members were undecided, and Rankin tried to convert them to his point of view: "They have demoralized the people and misled them," they have rushed "into the Federal courts and trumped up every conceivable scheme to enjoin and prevent any kind of electric development, and if you begin lending to private corporations, real rural electrification will be paralyzed."[21] He offered an amendment to exclude the utilities.

Rayburn was well-aware of the hostility toward the power companies, for his own constituents had long complained about their arrogance. But he knew, too, that the House was in no mood to exclude them, and Rankin, though a friend of the farmer, was a real danger to the bill because REA opponents might agree to his amendment in order to kill the measure on the final vote. Moving swiftly, Rayburn sympathized with Rankin, agreeing that the industry had abused the market, but given the enormous task of electrifying six million farms, he wanted REA able to tap the resources of private enterprise if needed. "There are so many communities that cannot possibly qualify . . . much less ever pay themselves out and make a self-liquidating proposition, that these words utilities should stay in the bill."[22] With this delivery he persuaded his colleagues to trust his judgment, and Rankin's amendment was defeated, a critical move in the progress of the measure.

Having passed the House with the two principal amendments on interest rates and utility participation, the bill went to the conference committee. Norris was the only senator to attend regularly, but Rayburn was usually accompanied by Carl Mapes of Michigan and George Huddleston of Alabama. Rankin also attended the meetings as an observer. Norris described the conference committee as the place where "the real battle developed. I had a stubborn, embittered fight on my hands" because the "House conferees were most insistent upon retention of the House amendments to the Senate bill."[23] The senator was equally determined that the House amendments not become law, and a deadlock developed. "We quarreled for a long time," Norris later recalled, but "no one lost his temper . . . when the Conference was over each time, it adjourned and they were all friendly with each other."[24] Neither side would retreat, however, and REA was in danger of dying in conference.

Norris announced that further meetings were futile, that the committee was wasting its time and that he was going to make rural electrification wholly as a public enterprise an issue in the upcoming 1936 election. With these words he stormed out of the room, but Rayburn quickly followed him and said, "Now Senator, don't be discouraged . . . we will come together because we have made up our minds you are not going to give up." Trying to soothe Norris, Rayburn continued: "Just let it rest awhile . . . within a few days we will notify you we are ready to have another meeting."[25] Rankin also tried to calm the senator, and the two, who agreed philosophically that electric companies should be excluded, reviewed the situation. Rankin asked his friend to give the House conferees more time, to wait a few days for new developments. Norris agreed.

At the root of the difference over the interest rate was the question that has longed haunted REA: should public rural electrification be subsidized? The power industry had taken the position that all farms able to use service without subsidy were already served, a position reiterated at the July meeting with Cooke in 1935. Cooke had insisted since his days with the Giant Power Board that if area coverage, graduated rates, long-term loans and special loans for wiring and appliances were furnished, rural service could be self-supporting. Indications that Cooke's theory had merit were available; power companies had lowered rates in some areas where REA work had already started. Still the subsidy question could not be ignored. Only a few months earlier Cooke had discussed it with members of his staff, and when the REA administrator appeared before the Committee on Interstate and Foreign Commerce, he had admitted that subsidy might be necessary to serve the poorest farms.[26]

Norris, Rankin, Judson King and other ardent proponents of REA had little difficulty justifying the subsidy. Since 3 percent was the "going rate" of interest in 1936, Norris did not see his proposal to anchor REA rates at the same figure as a step toward subsidization. In the event that rates went up, however, he thought REA deserved the advantage: "I believe we would be entirely justified in providing for a subsidy for rural electrification," he told a colleague, because "the matter is of such vital importance."[27] House conferees were convinced, nonetheless, that a fixed rate of interest would eventually lead to subsidization, for rates were sure to increase.

Meanwhile Cooke was watching the battle and realized REA was in jeopardy. Careful not to offend Norris or Rayburn, he politely suggested to the latter that the rate of interest "be no more than the average rate of interest payable by the United States of America on its obligations," which at the time was 3 percent.[28] Such a formula left the door open for changes in rates, and it satisfied Norris's demand for 3 percent. Cooke told Norris the question of loans to private companies was academic because REA would require area coverage to qualify for any loan, something the electric companies would not agree to. After conferring with both parties, Cooke informed the president that the controversy was "likely to be resolved." Norris expected to compromise, for he told a colleague that when he drafted the bill, "I had gone as far as it was possible."[29]

When the conferees returned to their negotiations a mood of compromise was apparent. As soon as Rayburn brought up Cooke's suggestion about the rate of interest, Norris assented. On the matter of loans to utilities, Norris agreed to let "persons and corporations" be eligible, but only after preference had been given to public bodies, meaning cooperatives. The Senate and House accepted the measure as reported out of conference, and Congress passed the bill on May 11, 1936, exactly one year after the president's executive order had created REA. On May 21, Roosevelt signed the bill, and on May 26 the Senate confirmed Morris Cooke as administrator of the new agency.[30]

The measure converted the emergency relief REA into an independent agency. Loans would be made principally to cooperatives on a self-liquidating basis with a twenty-five year amortization period. Families needing money for wiring and appliances could apply for small individual loans. For each of the first two years, RFC would supply REA with $50,000,000, and for the next eight years, Congress was authorized to appropriate funds up to $40,000,000 per year. REA, therefore, was given a ten-year life span.[31]

In regard to the expenditure of funds in the states, provision was made that "the annual sums available for loans will be allocated yearly to the total of such unelectrified farms in the United States."[32] In other words, a state such as Georgia with 2.8 percent electrification would receive more funds than California with 63.9 percent electrification. The schedule for the first year of operation showed the former state receiving $1,010,250, and the latter $287,500. Poorer states such as those in the South were given a distinct advantage, but there is no evidence to suggest that southern legislators were responsible for that provision. The purpose was to attack the problem where it was greatest.[33]

REA had a distinct advantage over the electric companies in the all-important matter of interest rates which, according to the agreement made in the conference committee, would be the same as that on long-term federal securities. In 1936 that figure was 3 percent, but the interest yield on utility securities for 1935-36 varied from 3.25 to 3.88 percent. Cooke had informed Norris of the higher cost to the private corporations which likely caused the senator to accept the House amendment. Composite average interest rates for the years through 1944 show public utility bonds varying from a high of 3.93 in 1937 to a low of 2.97 in 1944. REA rates for the same period varied from a high of 2.88 percent in 1938 to a low of 2.46 percent in 1941.[34]

The new REA went into operation July 1, 1936, with the new fiscal year. Supporters of the agency were relieved that the rigors of the past year were over; congressional differences of opinion over power company participation nearly killed the agency. In the move to give REA permanent status, Rayburn's role was secondary to that of Norris, although Cooke later recalled that Rayburn deserved equal recognition for his action. Rayburn's contributions to rural electrification would acquire greater importance in the future. The immediate task was the organization of cooperatives to serve the public.[35]

7
ELECTRICITY
COMES TO
THE FARM

Almost as soon as Congress finished with the REA measure in 1936, Cooke looked for a successor. His departure was characteristic; he was an innovator, not an administrator. In his search, Cooke kept in mind that for the next few years REA would experience speedy growth, since it had a large backlog of requests from farmers for action, but he anticipated continued resistance from the power companies and their supportive apparatus. He needed someone able to withstand pressure from the electrical industry as well as organize the agency's administrative details. Cooke found his man in John M. Carmody who established a successful record, tripling the number of electrified farms during his tenure at REA.

Carmody had more than a quarter-century of experience in industries such as steel, coal and garments. He edited *Coal Age* and *Factory and Industrial Management* for six years, and in 1931 had made a field survey of industrial developments in the Soviet Union for McGraw-Hill. His government experience began in 1922 when he directed a study for the United States Coal Commission. In 1933 he served as chairman of the Bituminous Coal Labor Board; he had two years service with the National Mediation Board and had worked for the National Labor Relations Board when he joined REA. Although he had only limited experience with rural electrification, which he acquired as chief engineer of the Civil Works Administration, "he had a reputation as an efficient, no-nonsense administrator who," wrote one reporter, "excelled in getting the maximum out of an established bureaucracy."[1]

Carmody was appointed deputy administrator first, effective August 1, 1936, but he was the real head of the agency since Cooke was preoccupied as Roosevelt's "trouble-shooter." The president appointed him chairman of the Great Plains Drought Area Committee; he continued to serve on the National Power Policy Committee; and he was busy with the Third World Power Conference, a meeting of international experts on electrical energy held in Washington in September 1936. The president had twice refused to accept his resignation from REA, but in January 1937, Cooke informed Roosevelt he was leaving. "The REA is now a seasoned organization fully competent to carry on," he said, and Roosevelt accepted his resignation.[2] Carmody was named administrator of REA, a post he had in effect held since he joined the staff. Cooke's departure marked the end of his official ties with REA, for he deliberately avoided becoming entangled in its affairs. His last government post came in 1950 as chairman of President Truman's Water Resource Policy Commission, but he maintained an active interest in public affairs until his death in 1960.

When Carmody took over in August, REA was swamped with applications for service and was moving at a snail's pace. "It was almost dead," Carmody later recalled, "when I went there." He ordered the information service to stop releasing stories of "marvelous progress." "The few projects that had been started," he added, "were small and creeping. . . ."[3] The new administrator requested $1,000,000 from RFC, the first funds obtained under the new law. Under Carmody's prodding the momentum picked up, and in January he called for another $15,000,000.

Cooperatives were usually founded when farmers and townspeople met and agreed to apply to REA for a loan. In some cases, REA field workers furnished assistance, but their advice was low-key, although by the latter part of the decade this promotional work was quite overt. Local officers of the Grange or the Farm Bureau were especially active in the establishment of co-ops. Congressmen lent a helping hand, serving as liaisons between their constituents and the agency. Rural electrification was popular among congressmen, for they could easily present themselves as the initiators of REA services in their districts. Despite their sometimes taking too much credit, they helped the program. They organized meetings, gave advice on legal and financial affairs, speeded up the processing of applications, and in general were catalytic agents in the bureaucratic rigors of obtaining REA service.[4]

Once local officers of the co-op were selected, they canvassed the area for potential recipients of service. Care was taken to include as many customers as possible to provide area coverage, the basis for successful operation of the co-op. Members could not live in areas already served by a power company, or within the boundaries of any city, village or borough having a population in excess of 1,500 inhabitants. Only a $5.00 deposit was usually required.[5]

Cooperatives and REA had a working partnership. In applying for a loan the former had to provide sufficient information to enable REA officers to determine its likelihood of success. This information included maps of proposed lines that had to be arranged to provide area coverage. Price schedules with graduated rates were required, even though the energy itself would be purchased from power companies. Solvency of the co-op, something with which the Washington agency was particularly sensitive, was based on the likelihood of each co-op meeting not only its loan payments but also the cost of maintenance, insurance, payroll, depreciation and other expenses. Each co-op was required to maintain a reserve fund. REA officers furnished supervision and assistance and made the final decision on location of the distribution lines. They also had the right to approve the co-op's manager. In some instances the agency selected the site of the office and equipment yards. REA audited the co-op's accounts each year. Seldom did the two clash for they had a mutual understanding of the purpose of their work.[6]

It fell to the REA research section to lower construction costs, long one of the principal barriers to development. Cooperatives made group purchases of materials to get discounts from manufacturers. A special cyclometer was developed that enabled consumers to read their own meter and mail in their monthly payment. Special low-cost 600 VA transformers were used with those homes consuming small amounts of electricity. Prefabricated pole-top assemblies also reduced time and labor. Such shortcuts reduced the line cost to an average of $970 per mile, a real saving compared to the $2,000 figure the electric companies cited.[7]

Progress in connecting farms was rapid. Requests for assistance in establishing cooperatives were numerous, and REA approved loans as rapidly as funds and circumstances would permit. By the end of 1938, more than 350 projects were established in forty-five states, serving 1,450,000 farms. From the administrator to the laborer, enthusiasm overflowed. Their motto was: "If you put a light on every farm, you put a light in every heart."[8]

It was uniform practice for families to use electricity at first for three purposes: lights, an iron and a radio. Costlier conveniences such as running water and indoor bathrooms came later, sometimes with the aid of a small consumer loan from REA. Regional variations in the acquisition of appliances were evident: southerners acquired refrigerators at a higher rate than co-op members in the north central states. The converse was true of washing machines; northern recipients of REA service had a 75 percent saturation in that appliance in 1939 compared with a 21 percentage mark for the South. Families praised the new comforts, and one cooperative staged a mock funeral for a kerosene lantern to signify the end of life without modern conveniences.[9]

A particular problem of REA was the large number of indigent families simply too poor to subscribe to a cooperative. Impoverished farms were most numerous in the South, and for them REA devised the "Arkansas plan." Designed for one- to three-room houses, the standard $5.00 membership fee was paid with a first payment of 20 cents and monthly payments of 10 cents. An absolute minimum of $1.00 per month was charged for the first eleven kilowatt-hours, but rates for additional power were lower. For $10.00, REA would wire a two-room house if the tenant made a down payment of $1.00. An iron and radio were available from REA at a cost of $3.00 and $7.00 respectively. These payments were made in installments along with the payment for wiring, coming to a total of 79 cents per month. For the minimum usage of electricity, including the iron and radio, the monthly charge was $1.89.[10]

Penniless farmers were permitted to work as laborers on REA crews in nearby areas as a form of "payment in kind." Families at the bottom of the agricultural ladder, therefore, could have some benefits of modern living, and REA reported the Arkansas plan had "been adopted mostly by systems in the southern states," the region crowded with the most penurious class of all.[11]

Power companies continued, however, to bother REA. They found an effective weapon in the legal and bureaucratic procedure involved in the organization of a cooperative. To begin with, lawyers often refused to serve as consultants for the co-ops because they were either on retainer of power companies or were afraid of opposing them. State public utility commissions were another stumbling block since they often represented the private interests. In Massachusetts, a state particularly hostile to REA, the utility commission refused to grant articles of incorporation to the

Tri-County Project which consisted of 700 families "not one of whom," Carmody claimed, "was being served by any existing utility or had any promise of service except through REA."[12]

Some state commissioners ruled that cooperatives could not build lines within one mile of existing company lines, a rule that often prevented area coverage. REA fought such rulings, and Murray Lincoln of the Ohio Farm Bureau thought cooperatives should not be subject to state commissions because they were not competing either actually or potentially with corporations organized for profit. Power companies challenged co-ops in court either in regard to rates or franchise rights. Not until the United States Supreme Court ruled favorably for REA in several cases involving these questions did legal harassment stop. In the meantime, harassment had served its purpose: to delay the co-ops and give the companies time to build their own lines into the areas in question and preempt the co-ops regardless of the Court's ruling.[13]

Even state rural electric bureaus, created in order to assist the Washington agency, catered to the electrical industry. With few exceptions they were initiated or seized by power companies, and their real intent was to thwart REA. For this reason, REA avoided them and worked directly with the cooperatives.[14]

Even county agents of the Agricultural Extension Service and professors of the agricultural schools fought the agency. County agents tried to frighten and discourage farmers from joining co-ops, saying that membership would make them liable for dangerous risks and that they would be held personally accountable for legal suits. REA service, they added, would not be reliable because shortcuts in construction of the lines would be taken. A few county agents participated in the organization of co-ops, only to turn the maps and rate schedules over to power companies which quickly built into the most lucrative areas in order to kill the co-ops. When REA had temporary status, Cooke had warned about "college professors acting as undercover representatives of private industry."[15] One Virginia Grange officer reported that a professor at the agricultural college "was going over the state advising farmers against cooperatives, and advising that they assist the power companies . . . because of certain restrictions imposed by the REA."[16]

The situation reflected the complex set of circumstances standing in the way of REA. It demonstrated that USDA was not free of influence against public power. Although many county agents worked on behalf of REA, others were beholden to power companies since in some states the latter paid a portion of their salaries. Professors of agricultural engineering also took a partisan stance because they placed a portion of their graduates in the electrical industry. Professors benefited from grants, promotional publicity and lucrative employment during sabbaticals or leaves of absence, all or part frequently furnished by one or more of the utilities operating in the state. Although extension agents were federal employees, they had a comradery with their professors and classmates who had chosen employment with the electric companies. A sense of brotherhood existed, and they had a common allegiance to their alma mater. They tended to hold similar views on public power and other socioeconomic questions of the day, so that extension service resistance to REA was strongest in those states where the colleges and the industry had close ties.[17]

However menacing legal harassment or saboteurs might be, the most serious damage to REA came in the form of the "spite lines" that power companies extended into lucrative sections of the countryside, leaving cooperatives with only the poorest farms and homes. This practice endangered a co-op's solvency. If local inhabitants preferred private service, then REA stayed out of the area and welcomed bona fide electrification privately administered, but spite lines were another matter.

In Louisiana, a power company ran extension lines in a pattern resembling the spokes of a wheel without connecting the homes between each line. The respective cooperative could do nothing because it was left with too few and too widely separated customers to be solvent. Private lines were often run into an uninhabited area in order to kill any opportunity for a co-op to serve it in the future. For the most part, however, spite lines served only customers yielding high revenues, a serious impairment of the cooperative's ability to provide area coverage since revenues from the prosperous farms were needed to offset the losses incurred in serving the poorer ones.

Normally this "skimming the cream" occurred only after the residents had announced their intention to establish a cooperative. Whenever possible, REA negotiated with the respective companies, sometimes purchasing their lines, but Carmody on occasion had to meet particularly obstinate companies on their own terms. In one instance he received a tip that company work crews planned to start construction at midnight Sunday to break up a co-op in Michigan. He ordered the REA crew out at midnight Saturday. "It served notice," he said, "in a way they understood."[18]

Supporters of REA questioned why power companies took a sudden interest in the rural market after ignoring it for so long. How could they find reason to build lines, it was asked, when only a short time before the industry had reported to REA that "there are very few farms requiring electric service. . . ."[19] Power companies had no answer, except to say that general prosperity had improved to the point that rural service was profitable. But critics of the industry disagreed; as one reported, "it became obvious to the power companies that in spite of all they had said to the contrary, the farm market had great potentialities."[20]

Not all private construction came in the form of spite lines. Beginning in 1936, many power companies extended service to farmers in a legitimate manner and did not require them to pay the construction costs. Rates were also comparable to those of the cooperatives. This belated entry into the market was due to REA competition and the realization that the agricultural market was lucrative. In 1939 power companies still served the largest portion of electrified farms, but that included dwellings already connected in 1936. Once REA went into operation, it connected farms at a faster pace than the utilities, a practice that continued until electrification in the United States was complete. A handful of power companies borrowed REA funds, but they accounted for less than 2 percent of the loans.[21]

Rates varied from area to area for both cooperatives and utility companies because of population density, amount of consumption and the presence of a low-cost public supply of energy such as that generated by TVA. Cooperatives supplied by the latter, approximately 27 percent of the total in 1941, enjoyed the lowest rates in the United States, ranging from 2 to 3 cents per kilowatt-hour. But most co-ops were not so fortunate and they had to buy power wholesale from private suppliers. Customers served by these co-ops had to pay about 3 to 5 cents per kilowatt-hour.[22]

Rates for homes served by utility companies were comparable, but conditions gave the utilities a competitive edge. They generally served only the lucrative farms, those with high usage such as dairy farms and those where the population density was greatest. At the same time, private firms could offset a portion of their rural costs with revenue gained from serving towns and cities, an opportunity closed by the REA Act to cooperatives. Thus, rates for the latter were assessed under costlier conditions. Cooperatives tried to provide the first 100 kilowatt-hours each month at the lowest rate and recapture their cost as recipients increased their use of electricity. Such varied circumstances made it impossible to determine, according to one study, "whether cooperative rates were higher or lower than those of private utilities serving similar territory."[23] But rates were remarkably lower than those in use prior to the creation of REA, indicating that a significant accomplishment of the agency was its effect in getting the companies not only to serve rural areas but at reduced prices.

TABLE 1:   Average Net Monthly Bills of Private and Municipal Utilities and Rural Cooperatives for Rural Electric Service in Wisconsin Communities, July 1, 1939

Monthly Use (KWH)	Private	Municipal	Cooperative
40	$3.88	$4.03	$3.49
50	4.02	4.21	3.92
100	5.61	5.33	5.83
150	6.88	4.71	6.99
250	8.40	8.85	8.91

SOURCE: The Twentieth Century Fund, *Electric Power and Government Policy* (New York, 1948), p. 466.

Whatever the differences over rates, local response to the use of electricity was overwhelming. Rural folk needed no instructions on the benefits and use of electricity because they had seen it utilized in towns and cities for a generation. When offered the chance to join a co-op, they responded enthusiastically. They also responded to REA educational and promotional efforts such as the home demonstration program. Women employees of the agency conducted "kitchen parties" on the use of appliances, cooking with electricity, refrigeration and other applications of electricity in housekeeping. Appliance sales skyrocketed in towns and cities when REA built lines into the nearby rural environs. An officer of the Arkansas Farm Bureau commented that "electricity at first was regarded as a luxury by rural people. Almost overnight it brought to them the conveniences about which they had dreamed for a long time."[24]

Expressing the relief from drudgery is an excerpt from a poem a housewife wrote about the impact of REA in her life.

> When you and I were seventeen, it was
>   different on the farm . . .
> Sure we had running water, but we had
>   to run out to the pump with a
>   pail to get it.
> Blue Monday got its name from the way
>   the womenfolk felt at the end of
>   the washday.
> We didn't have radio, electricity is
>   required to power them.
> Now things are different—since the
>   high line went in.
> There is just as much to do on the farm
>   as ever, but it is a lot easier to
>   do it.[25]

By 1939 the improvements wrought by electricity were visible in rural life. REA had 417 cooperatives serving 268,000 households and had loaned $3,644,711 for wiring and plumbing. About 25 percent of all farms had service. Carmody had molded REA into an efficient and smooth-ly-operating organization known for fast work. Morale was high as evidenced by the employees' willingness to work extra hours. Successful in its purpose and free of interference, REA was widely admired for improving rural life. But the rapid growth and high morale soon ended for the agency was about to experience a dramatic change in structure and leadership.

8
THE SLATTERY
INCIDENT—1939-1944

From 1939 through 1944, REA went through a tumultuous period.
Several electric companies tried to discredit the agency in hopes that
Congress would reduce its role when World War II ended. More significant,
however, was the "Slattery incident," an internecine war among Harry
Slattery, a new REA administrator, his staff and Secretary of Agriculture
Claude R. Wickard. Closely related to these events was the creation of
the National Rural Electric Cooperative Association (NRECA), a private
lobby seeking to expand public electrification, but which differed sharply
with Slattery over policy critical to connecting the remaining 4,000,000
farms without service. Charges and countercharges were made by the
participants until the Senate conducted an investigation into the incident
in 1944 and drew the president into the fracas.

The roots of this hectic era went back to 1939 when the independent
REA was transferred to USDA. Carmody thought the move would end
the autonomy of the agency and resigned. Harry Slattery, undersecretary
of the interior, succeeded him and his tenure at REA was marked by
strife and discontent. An omen of things to come was evident when
George Norris told Roosevelt: "the recent order transferring this . . .
organization is going to result in great damage."[1]

Slattery stepped into a cauldron boiling with jealousy and suspicion.
For two months Roosevelt had searched for Carmody's replacement, and
some REA division chiefs, particularly acting administrator Robert Craig,

had campaigned for the post. With Slattery's appointment they felt cheated, and although acknowledging the prerogatives of Slattery's office, they gave him no loyalty. On the contrary, they cultivated and maintained their own followings.[2]

Confusion and disagreement over administrative procedure also existed between REA and USDA from the moment of the transfer. A special inter-departmental liaison committee had criticized REA personnel practices, namely, the larger than usual non-civil service appointments, most of which were for technical positions. When the agency was created in 1935, Cooke had to find competent technicians where he could since most were employed by or dependent upon the electrical industry. Carmody had continued making such appointments until USDA chief Henry Wallace told Slattery that REA "was organized on a lopsided basis."[3]

Personnel appointments were the only real difference Slattery had with USDA during his first year. Operations proceeded smoothly; another 87,080 miles of line were strung.[4] Slattery drew on his long association with Wallace who did not want to disturb REA. The real drama of the "incident" and the events leading to the creation of the NRECA began when Wallace resigned in 1940 to run for vice president and Claude R. Wickard took his place.[5]

Born in 1893 to devotees of the United Brethren Church, Wickard grew up on the family-owned Fairacre Farms in Camden, Indiana. Hard work, discipline and a genuine love of the soil were instilled into him early in life, and he had a knack for politics acquired from his grand-father. After graduating from Purdue in 1915, he managed the farm, was active in the Indiana Farm Bureau and was regarded throughout the state as a competent and realistic agriculturist with "grime in his knuckles." In 1932 he was elected to the Indiana state Senate, but politics were peripheral to Wickard's interest, and in 1933 he went to work in the AAA Corn-Hog Section. Professionally competent, a midwestern Demo-crat and admired by Wallace, he moved quickly into more responsible posts, becoming head of the Corn-Hog Section in 1935, moving to undersecretary of agriculture later in the same year Wallace resigned. A man known for his "set ways," the new secretary proceeded to remake USDA to his own liking.[6]

Slattery and Wickard first clashed over the administrative handling of the cooperatives. REA had always taken pride in viewing them as private

corporations, operated according to sound business principles, highly successful and, best of all, rooted in the local community. The agency granted
as much autonomy to the cooperatives as possible, and when Wickard
stated they were not private institutions, but semi-private and that he intended to have a hand in their affairs, Slattery was upset. Debate over the
public or private character of the cooperative was academic, and Slattery's
reaction was unwarranted since the co-ops were semipublic. But the old-
line REA supporters such as Slattery were conditioned by their long
battle with the power industry; the latter had always insisted that electrification was possible only with full public support. Slattery and many
others believed that it could be accomplished with limited public support,
that the quality of private enterprise could be maintained. Hence, their
insistence, not wholly incorrect, that cooperatives were private institutions. Wickard, nonetheless, requested a list of all cooperative officers
and directors and indicated he would hence pass on the selection of
superintendents. When the new secretary also announced that "the time
has come for full integration" of REA into USDA and appointed a special
"integration committee" to carry out that objective, Slattery insisted
that his real purpose was to oversee selection of co-op officers.[7]

Wickard next created a new post of second deputy administrator in
REA and appointed a new personnel director. Appointment of the latter
was arranged to coincide with an investigation of REA personnel. A special
team, whom Slattery referred to as "sleuths," looked into the internal
affairs of the agency. This move by Wickard backfired, however, when
columnist Drew Pearson published an account of the probe, saying the
searchers were pro-utility men formerly employed in the defunct Public
Works Administration. Angry and embarrassed, Wickard accused Slattery
of inspiring the publication. The REA administrator took delight in his
chief's embarrassment, since he interpreted the probe as an attempt to
circumvent his authority.[8]

Slattery was accurate in his assessment of Wickard's purposes; Wickard
fully intended, as stated by his assistant Samuel Bledsoe, "to move him
[Slattery] where he would not be in the way."[9] Wickard was convinced
that Slattery was injurious to rural electrification and in a broader sense
to USDA. The secretary and his assistant Paul Appleby clandestinely
visited Senator Norris and Congressman Rankin to secure support for removing Slattery. They alleged that REA was in fiscal trouble and near

collapse. Neither Norris nor Rankin accepted the charges, and Norris told Wickard not to listen "to a lot of gossip of men in women's skirts. . . ."[10] The two members of Congress, if anything, were angry and regarded their visitors as intruders into an agency known for its independence and its achievements in improving rural life. Disturbed, Norris and Rankin urged the president to make REA independent again, but Roosevelt only ordered the Bureau of the Budget to examine the fiscal affairs of REA. He told Rankin, "I hope this will work out satisfactorily to Claude Wickard and Harry."[11] No evidence of fiscal mismanagement came to light.

Considering that Wickard had worked with Slattery for less than a year, his effort to oust the administrator might appear hasty. Slattery accused him of political chicanery. Wickard probably did not fully understand REA's method of operation and interpreted them as a challenge to his authority. That alone, however, would not explain his conviction that Slattery must go.

Wickard's "sleuths" exposed the favorable treatment that certain REA division chiefs gave the copper industry. Carmody had followed the general rule of using 60 percent aluminum conductors, the cheapest, and 40 percent copper, slightly more efficient in some cases. Furthermore, he had instructed the cooperatives to use competitive bidding in awarding contracts for conductors. During the two-month interval between Carmody and Slattery, however, the copper interests, the Copper Wire Engineering Association (CWEA), a representative of the few major suppliers of copper wire, had managed to ingratiate itself with a handful of department heads in the engineering division, the "copper crowd." They reversed the proportion to 60 percent copper, 40 percent aluminum, and stopped the practice of competitive bidding. In some cases copper companies knew of new construction allotments as soon as the co-op superintendent and thus were able to obtain orders before their competitors had a chance to bid. Slattery restored competitive bidding as an option, but the engineering division would delay approval of projects in order to sabotage competition. As a result, Slattery's order had no real effect. "No one knows how much damage they did," Carmody remarked about his own experience with this problem, "until I caught up with them."[12] Deputy administrator Craig favored the use of copper and with the advantage of Slattery's frequent absences due to ill health, exerted his influence on its behalf. CWEA had also preferred Craig as REA administrator.

Internal dissension worsened and influence peddling on behalf of copper surfaced with another development, the resignation of Boyd Fisher. Angry over a variety of matters, including the machinations of the copper crowd, Fisher left and started a one-man crusade to cleanse REA. He drew up a sixty-six page explanation of the intrigue, influence-peddling and other injurious activities that increasingly characterized the agency. He mailed 400 copies of his rendition to newspapers, magazines and government officials, portraying Wickard as a petty politician and Appleby and Craig as tools of special interests. He referred to CWEA as "the most effective sales organization I have ever encountered." At the root of the trouble, Fisher insisted, was Slattery who had been too indecisive, too timid with opposition, and too inclined to "delay and discuss at great length." The administrator had "in all important respects, completely cancelled out."[13]

Fisher had an explanation in accord with Wickard's. Praising Slattery's record of "warfare in behalf of the public good," Fisher concluded, nonetheless, that something was "out of the ordinary." For one thing, he stated, "Mr. Slattery has been absent from REA, because of ill health, for nearly three months at a time." So frequent were the absences that he had not been "on the job long enough . . . to pull the threads of management out of Robert Craig's control." Since few staff members had met the new administrator or been able to find him, they went to Craig for decisions. Still, Fisher added, Slattery's absenteeism did not account for his day-to-day inconsistency, his self-contradictions, and his almost pathological suspicion of those around him. "One can only conclude," wrote Fisher, "that since he took this job he has not been himself."[14] Wickard had come to the same conclusion and, according to Bledsoe, "He [Slattery] knew he was slipping and apparently had a persecution complex and needed needed extraordinary methods to keep his control and prestige in REA."[15]

Shortly after assuming his duties at REA, Slattery had been hospitalized in Miami for three months with pneumonia. Less serious illnesses accounted for his shorter, but frequent absences from his desk. One cooperative officer described him as "nervous and twitchy," ill-at-ease, in another world even in the company of others. Wickard's assistant Bledsoe later recalled: "Slattery was in declining health and was actually senile. I never felt he was quite himself."[16] David Lilienthal, a friend of Slattery, wrote: "However pretty a picture they seek to put on it, the unhappy fact of the business is that Slattery is a sick man, of little force, and yet highly deserving as an old-time progressive."[17]

Wickard relentlessly maintained the pressure. Slattery tried to fight back by building a following of his own, and whenever possible he appointed old friends to REA slots, usually as non-civil service consultants, which only reinforced his opponents' charge that he refused to follow USDA procedures. A wise appointment was Judson King as his special consultant. King belonged to the old-line public power camp, and his reputation gave credence to REA not only as a consumer-minded agency but also one with experienced officers. But King soon became embroiled in the dispute through the REA labor policy.

Cooperatives used non-union labor because it was cheaper, and also because union linemen and electricians were not always available. Job hungry farm boys often worked, therefore, as journeymen electricians. The International Brotherhood of Electrical Workers (IBEW) complained about the lower wages and inadequately trained personnel. In as much as the Roosevelt administration gave aid and sympathy to the working man, REA had to acknowledge the IBEW. Even more pressing, however, was the need for cooperatives to hold down costs, for many tottered on the brink of collapse during their first years of operation. Slattery and King agreed to arrange a compromise with IBEW.[18]

Since REA was not a profit-making organization, IBEW promised not to instigate strikes and to establish an educational program designed to make the crewmen conscious of safety and craftsmanship. For its part, REA would urge the well-established cooperatives to pay prevailing union wage rates and it would urge the floundering co-ops to negotiate wages. King had devised this policy.[19]

The labor policy was only partly implemented. Wickard opposed the policy and removed Slattery's labor advisor, Dave Fleming. Co-op superintendents opposed collective bargaining, thereby reinforcing Wickard. IBEW protested to Slattery who was helpless, and King was driven into a defensive position alongside Slattery since the cooperatives opposed the policy. Craig, always anxious to usurp Slattery, opposed it as well.[20]

It was, thus, a troublesome set of conditions in the REA administration when the United States entered World War II. Wickard wanted Slattery removed. Because of the latter's indecisiveness and ill health, ambitious bureaucrats and special interests promoted themselves. Strong differences over the labor policy had erupted. Ironically, the program moved at a rapid pace; altogether 146,031 miles of line and 437,425 customers were added during Slattery's first two years of office. Only the phenomenal rate of growth kept the cooperatives pleased and unconcerned with REA internal affairs.[21]

But cutbacks of REA material began in late 1941, forcing the agency to operate at a slower pace. Defense priorities also required that cooperatives extend service to new defense plants and new military installations such as army camps, flying fields, aviation schools and ammunition depots. Farmers accepted the restrictions as a patriotic sacrifice.[22]

In January 1942, however, a stringent system of priorities went into effect with the creation of the War Production Board (WPB). It stipulated that any project 40 percent complete as of December 5, 1941, would be permitted to finish. Projects not reaching the 40 percent mark prior to that date, plus all new or proposed projects, were stopped except those directly related to the war effort or considered essential to public health. Farmers hoping to receive service were bitterly disappointed, but the situation worsened when in August 1942, the WPB rescinded the 40 percent rule and placed a "freeze" on all construction of rural lines.[23]

Dissatisfaction with the new ruling was widespread, especially where farmers were on the verge of getting service. Many families had gone to the expense of wiring their homes and purchasing appliances. Compounding the problem from the farmers' standpoint was the increasing number of laborers going into military service or defense plants. Electrical energy, it was thought, was the best replacement for the lost farm labor. Complaints were numerous. "There are many hundreds of farm families," one co-op superintendent pleaded with his congressman, "who look forward with anticipation to the day when their project is complete and would greatly appreciate your efforts in bringing about a release from the 'stop work order' that would permit its completion."[24]

Co-op officers were irritated, furthermore, because electric companies were able to acquire material denied to them. From the project superintendent to the REA administrator, it was charged that WPB was staffed with "dollar-a-year" utility executives taking advantage of the war to thwart REA. For nearly three months, Slattery insisted that "not one pound of copper went to the REA co-ops, although the utilities got their lion's share."[25] At the local level the same complaint was heard: "since the Texas Power and Light Company is able to obtain the necessary materials for new construction," protested one superintendent, "we would like to have the assurance from OPM [Office of Production Management] that we would receive the wire necessary for the completion of our project."[26] To what extent REA suffered from WPB discrimination is impossible to determine because the electrical industry had the advantage of wartime priority. To obtain more lenient consideration, REA launched, wrote Lilienthal, one "of the most remarkable pressure campaigns by one branch of the Government against another that I have ever heard of."[27] Although the WPB may not have merited such an attack, the double standard of supplying cooperatives and power companies suggested some special interest influence.

Two incidents occurring simultaneously in the South symbolized the clash over public and private development of rural electrification: the construction of an REA transmission line to feed an Arkansas aluminum plant and the alleged REA hoarding of copper in Texas. The Defense Plant Corporation, a wartime bureau and builder of the plant, had awarded the energy contract to the Ark-La Cooperative of Homer, Louisiana, which proposed to construct a line from the Grand River dam in Oklahoma to Lake Catherine, Arkansas, site of the aluminum plant. The Arkansas Power and Light Company hoped to win the lucrative contract, but lost when Slattery approved a $2,500,000 loan to the cooperative for the project.

A short time later Kansas Representative Thomas Winter, regarded by the REA group as a spokesman for the power interests, accused REA of wasting strategic materials, thereby promoting the "socialization of the electrical industry." The Faddis Special Committee of the House Military Affairs Committee investigated the incident and concluded that REA would waste materials in supplying the plant because the proposed line duplicated the private lines already in existence. Congressman Rankin and other friends of REA protested the subcommittee report, but the full committee on military affairs overwhelmingly approved it. President Roosevelt publicly declared the investigation biased in favor of the utilities, but otherwise did not become involved. Congress took no further action, and the cooperative built the line. However, Congress trimmed the REA budget for the next year beyond the ordinary cutbacks associated with the war, doubtlessly a reaction to the charges of waste. *New York Times* columnist Arthur Krock expressed the mood of the lawmakers: "Congress did not appropriate REA money to build transmission lines to great industrial plants, but to bring comfort to and relieve drudgery on the farm."[28]

REA was also charged with hoarding large quantities of copper in vacant cotton fields in east Texas when the armed forces were short of this critical resource. Again it was Winter who charged on the floor of the House in December 1941 that he had photographic evidence of the hidden copper, that REA was "teeming with Communists, fellow travelers and bureaucrats" whose ultimate goal was the socialization of the electrical industry. The same Faddis committee, unfriendly to public power, investigated the alleged hoarding. Before committee questioning, Secretary Wickard defended the agency, including its approval of the Ark-La line; Slattery said the copper was in storage until future use and Arkansas Congressman Clyde T. Ellis, a staunch supporter of REA, reported that a private contractor owned the wire and could not sell it since the WPB freeze went into effect.[29]

The hoarding charges were resolved in favor of REA, but in the meantime critics of the agency made use of Winter's accusations. Unfriendly newspapers reported the incident so it would appear detrimental to the agency; unfriendly members of Congress continued their assault and the trade press of the power industry stigmatized REA as a self-serving bureaucracy. Hoping to taint the image of the popular REA, the industry wanted to turn opinion against it and give utilities the opportunity to control the rural market at the war's end.[30]

Supporters of REA were afraid of future attacks on the agency and their concern for its protection convinced them of the need to organize a private lobby on its behalf. Discussion of an independent national organization dedicated to promotion of REA went as far back as Cooke's administration. Circumstances at the time were not appropriate for it, and the idea languished until 1939 when Carmody briefly entertained the possibility. Nothing was done, but his interest had advanced the notion that a "national" was needed. Well-established lobbyists had represented special interests for years, and why not, it was reasoned, have something of a similar nature to speak for rural electrification. The 800 or more cooperatives in operation could be represented as a single unit and apply considerable pressure on Congress for favors and especially larger appropriations.

For some time co-op superintendents had expressed need for decentralization of the program to provide help and closer supervision at the local level. "The cooperatives were learning the importance of a united force," wrote Harold Serverson of the Association of Illinois Electric Cooperatives. "Not until they banded together could their voice be heard with maximum effectiveness."[31] With the belligerence of the electrical industry, the idea of establishing a "national" acquired more importance every day. Squabbling and lack of decisive leadership in REA headquarters also made the cooperatives feel less secure and long for protection.

The hearings over the copper hoarding scandal served as the catalytic agent in the creation of NRECA for it was at that time, wrote Ellis, that public power enthusiasts, REA officers and co-op leaders from several states "discussed among themselves many common problems and agreed on the urgent need to establish a national association to speak for them and protect their interests in Washington."[32] On March 16, 1942, five prominent state co-op leaders held a preliminary meeting at the Netherland Plaza Hotel in Cincinnati to discuss organizational plans: William Jackson of New Jersey, Steve Tate of Georgia, William Sullivan of Tennessee, Thomas Fitzhugh of Arkansas and E. D. H. Farrow of Texas attended. Each had been involved in bitter fights with utility companies. They agreed to meet again in Washington in order to obtain advice and information from REA.

Recovening the next day at the Willard Hotel in Washington, they met with several REA officers. During the next few days bylaws were written, organizational plans were discussed and a tentative list of a board of directors was made. It was decided that the board of directors would represent each region of the United States, a practice patterned after the administrative organization of REA. On March 20, 1942, the articles of incorporation were filed in the District of Columbia, and the board of directors elected Steve Tate as president. That same day the newly-founded NRECA hosted a reception of several REA supporters in Congress, including George Norris, John Rankin, Robert Poage of Texas and others. Slattery was fully aware of these proceedings from the beginning and had participated to a limited extent in some of the discussions. A short time later he wrote Tate: "I want to pledge to you my effort and assistance that together we can achieve the great purpose of our President and the Congress of the United States when they created the REA."[33]

Headquartered in the District of Columbia, NRECA was dedicated to providing the remaining 5,000,000 unelectrified farms with service and obtaining lower wholesale rates of energy from power companies already supplying the cooperatives. It also wanted to protect them from the devices used by the utilities to thwart REA, such as influence with Congress and state agencies, legal suits and discriminatory taxes. Each cooperative was invited to join, paying a $10.00 fee. An annual fee of 25 cents was assessed for each family that belonged to the cooperative; the national headquarters received 10 cents and the remainder was left to develop subordinate NRECA organizations and programs in the respective state.

The NRECA was vigorous and grew rapidly. By January 1943, when it held its first annual meeting in St. Louis, 175 cooperatives had joined and many others were following suit. The association published a magazine, *Rural Electrification,* and lobbied strongly in Washington and in state capitals on behalf of public electrification. It assisted cooperatives with a variety of business services: safety and job training for employees, legislative research, government contracts, educational programs, legal advice and public relations. The Farm Bureau welcomed the new organization as a partner in the fight to supply rural inhabitants with the conveniences of modern life. Much of the success of the NRECA was due to the work of the first executive manager, Clyde T. Ellis.

For a leader, NRECA wanted a man able to meet several demanding qualifications: he must have championed public electrification and preferably have a rural background; he must have legal experience and also know the procedures and intricacies of Washington officialdom; he must be adept at public relations and acquainted with rural electric leaders. The nod had gone to Arkansas Congressman Clyde T. Ellis, defeated in a bid for the Senate in 1942.

As was characteristic of rural electric proponents, Ellis had grown up on a farm without modern conveniences. Born in Garfield, Arkansas, in 1909, a short distance from the site of the Civil War battle at Pea Ridge, his homeland was one of the South's poorest, where oxen were still in use on the eve of World War II. Electricity was not available, and Ellis's family lived with hand-pumped wells and outdoor privies, even though they observed life with electrical conveniences in the nearby towns. With a sense of bitterness, Ellis would tell the story of one city visitor who at Sunday dinner ridiculed his parents as "country-jakes" for their old-fashioned house. Driven to avenge this second-rate status, he was restless, constantly in turmoil, forever seeking a way to satisfy his social conscience for his own kind. "It was these memories," he recalled, "that later, as a Congressman and as general manager of NRECA, made me devote almost my total time and energy . . . to the rural electrification program."[34]

Convinced that private enterprise would never extend power into rural areas, Ellis saw government intervention as the answer. His idol was George Norris, and he viewed the Arkansas White River, as Norris did the Tennessee, as a natural source of energy. Ellis was elected in 1932 to the lower house of the Arkansas legislature on a platform of public power and rural electric development, a frankly liberal position in a conservative state. Ellis successfully ran in 1938 for the United States House of Representatives, again on a platform of public rural electric development.[35]

His congressional career was brief but impressive. He quickly made friends with Norris and Rankin; he won approval of the Norfork hydroelectric dam on the White River; he repeatedly introduced legislation proposing an Arkansas Valley Authority, and in 1942, he was designated chairman of the Southern Policy Committee, the contingent of southern congressmen devoted to the improvement of their region. That same year he ran for the Senate against John McClellan and lost. When NRECA was searching for a leader, Ellis was searching for new opportunities. He accepted an NRECA invitation to serve as the executive manager only on condition that the organization become the recognized spokesman for rural electrification. "I used the American Farm Bureau Federation," he later wrote, "as an example of the size and scope of organization I had in mind."[36]

Shortly after Ellis took over, NRECA persuaded WPB to rescind its freeze on REA construction projects. If a farmer could prove that use of electricity added to his production, REA was empowered to extend service to him. Slattery was pleased with the new rulings.[37]

NRECA could not, however, take full credit for the new rulings; REA, as noted, had stubbornly fought the WPB. Food production also had not reached the level considered essential for the war, and the Roosevelt administration wanted production increased. When NRECA pleaded for a modification of the rules, there was a natural inclination to listen since nearly everyone agreed that electrical energy was one of the best replacements of manpower. Still, as *Electrical World* noted, there was a connection between the lenient qualifications and Ellis's petitioning.[38]

In spite of their teamwork, Slattery and Ellis soon fell into a bitter feud over a NRECA plan to offer the cooperatives new and less expensive insurance coverage. Insurance was costly due to the hazardous work, the serious questions over workmen's skills and safety of the lines. Insurance companies were also convinced the cooperatives would eventually declare bankruptcy and fall prey to the utilities. By all indications Slattery approved the NRECA plan; Craig and REA insurance advisor Max Drefkoff had participated in the arrangements, and NRECA specially incorporated two insurance companies in January 1943.[39]

Slattery changed his mind and blocked the plan. He had heard that some NRECA members regarded him as an incompetent administrator, and although Ellis assured him that such talk was groundless, Slattery's suspicions were aroused. He urged the cooperatives not to commit themselves to the insurance offer because, he implied, NRECA's motives were questionable.[40]

NRECA officers pressured Slattery to reverse his decision. Ellis visited Roy Reid, USDA personnel chief, and asked him to relieve a list of "pensioners" from the REA payroll, "deadwood" whose duties allegedly did not justify their salaries and who were kept on only "because of their personal friendship with the Administrator."[41] Judson King was at the top of the list. That same day, April 24, Steve Tate contacted Slattery on the telephone and expressed surprise at his decision, saying "it is not consistent with what you told us." The REA chief replied that "I know more about how government funds can be used." Tate agreed, and then mentioned the "deadwood" on the REA staff. Slattery responded, "If you people are going to try to smoke me out, you've got a tough job." Tate ended the heated exchange with these words: "We're going to have a 'cat-thrashing.' "[42] Tate's prediction was accurate: Slattery and Ellis were headed for a direct clash, and each had a personality that would not permit compromise.

Slattery belonged to the first generation of conservationists and public power crusaders who were fiercely devoted to the use of federal authority. TVA and REA were two of their greatest achievements. Slattery had partly uncovered the violations that led to the Ballinger-Pinchot affair and also the Teapot Dome scandal; now he stood guard over REA with the zeal of the Spartans at Thermopylae. But Slattery had a suspicious nature that made him almost paranoid. He once threatened to check REA typewriters to stop leaks. These characteristics, a touch of paranoia and a quality of resistance intensified by lifelong conflict with special interests, caused him to see himself as the defender of an extraordinary agency brutally assaulted by Wickard and now Ellis.

Ellis was no less zealous or pugnacious. His life's mission was to end the misery and suffering of rural inhabitants living without the benefits of electricity. Impatient, occasionally belligerent, he belonged to the second generation of public power proponents wanting to go beyond the old guard into areas not perceived in the past to be within the confines of public enterprise. Quick to anger, he could be as vehement as Slattery. He disliked compromise and preferred not to wait for an opportune time; he figuratively plowed one furrow at a time and in a straight line. He and Slattery disagreed over the future role of REA.

In response to the pressure for approval of the insurance plan, Slattery fired Craig and his insurance advisor, Max Drefkoff. He was glad to be rid of Craig who still saw himself as the rightful heir to Carmody. Craig had campaigned for an organization such as NRECA since 1938. He and Ellis were friends, but the creation of NRECA and its insurance proposal were not related to Craig, though Slattery and King thought so. Upon his dismissal Craig accepted employment with the Copperweld Steel Company which had been involved in the copper scandal. The suspicious Slattery interpreted Craig's new job as proof of NRECA as a special interest seeking to gain control of REA.[43]

Slattery's inconsistent behavior and exaggerated suspicion convinced Ellis that Wickard's assessment of the REA administrator was correct. Now he viewed Slattery as an inept, contradictory bureaucrat who by default kept rural families from enjoying the comforts of modern life. Ellis shot off a request to Roosevelt for Slattery's dismissal on grounds of incompetence. Roosevelt refused.

Hoping to circumvent Slattery in another way, the NRECA executive fired a barrage of charges against him, fashionably called the "blitzkrieg." He alleged that REA was chaotic and in need of governance owing to a set of circumstances: the "pensioners" on the payroll, a general deterioration of administration as evidenced by a high rate of resignations, Slattery's frequent absences and a growing dissatisfaction in the public power camp with his performance. Slattery was in general said to be incompetent; it was rumored that he was insane.[44]

The USDA personnel department investigated the alleged "pensioners" and uncovered instances of unnecessary use of government funds for travel, some evidence of poor job performance, indications of confusion over responsibilities and a widespread feeling in the agency that certain employees were Slattery's spies. Still, the investigators concluded, the alleged "pensioners" were well-qualified and appeared to be conscientious workers. As to Slattery's repeated lengthy trips to Miami on behalf of the agency, frequently accompanied by some of the "pensioners," the investigators reported that it was difficult to conclude whether the trips were justified. NRECA insisted that so many long visits to the same place, particularly Miami, could have little to do with electrification. Morale among the employees was low, and Slattery attributed it and the resignations to the wartime transfer of the agency to St. Louis. It was unpopular with the staff, but they were right, nonetheless, to complain about the confusion and uncertainty in the agency, the lack of leadership, the lack of creativity and the disregard of instruction by some.[45]

Slattery replied with a volley of countercharges. He pointed out how the insurance proposal would permit unscrupulous manipulators to profit handsomely if they so desired. Ellis agreed and amended the plan, but it made no difference to Slattery. Rumor also spread that the construction companies building the power line to the aluminum plant in Arkansas had financed Ellis's campaign for the Senate, but no evidence substantiated the rumor. The real purpose of NRECA, Slattery insisted, was to seize REA for political benefit, to capture the four million votes of the families receiving REA service. Wickard and Ellis were allegedly as "thick as thieves." Judson King agreed and regularly referred to a Wickard-Ellis-Craig conspiracy. A figure as important as Senator Norris, just defeated for re-election in 1942, suspected a "sinister move" behind the attack on Slattery. If Ellis had a list of supporters and charges, so did the REA administrator.[46]

The extent of the division and confusion was evident when Morris Cooke politely nudged Slattery to step down so as not "to antagonize unduly the Administration in view of the war situation. We do have to make some concessions to power in the light of the war."[47] Cooke misjudged NRECA for it was anything but an instrument of the electrical industry. Gifford Pinchot made the same mistake, only he urged Slattery "to fight the thing through to the end."[48] Old-line progressives erred when they associated NRECA with the power industry; it was a fast-growing, certainly abrasive, child of the public power family. As the feud continued, it became clear that there were fundamental differences over the role of the REA; it amounted to a changing of the guard, new replacing the old.

Nowhere was this change more apparent than in the question of power generation by the cooperatives, a matter that since 1935 had remained quiet but persistent. Should the agency, it was asked, finance the construction of generating plants? Electric companies had always supplied the cooperatives at wholesale rates, but agreements were usually made only after strained negotiations. Wholesale rates, though not prohibitive, were still high, about 1.5 cents per kilowatt-hour. Prior to the war there had also been occasional shortages of energy which became acute with the pressing demands of defense. So serious were the shortages that "brown-outs" had occurred in rural areas. To provide an adequate and less costly supply, NRECA proposed that REA build generating plants, but others, including Slattery, disagreed. "He is firmly opposed to the tendency, which reaches into some parts of his own staff," wrote the editor of *The Public Utilities Report,* "growing up within the REA movement to expand their operations into the generating field."[49]

Slattery belonged to the older generation, including Cooke, who had never intended for REA to generate power, except in a minor way. Though the Rural Electrification Act authorized it, generation of power was not considered essential and frankly not within the confines of public rural electrification when the measure was passed in 1936. In opposing generation, Slattery only followed in the footsteps of his predecessors, but the power shortage and the anticipation of shortages after the war made the younger Ellis think otherwise.

The fight over power generation had already erupted in Texas where a dozen or more cooperatives were unhappy with the wholesale rates of the electric companies. Approximately 25 percent of the cooperatives' revenue went for wholesale power, and due to the technical nature of electricity, 25 percent was lost over the transmission from plant to consumer. Bolstering the voltage from 2,300 to 12,500 volts would solve the greater part of the latter problem. The companies refused, however, to provide the increased voltage or to lower wholesale rates. For two years negotiations had taken place when in November 1940 parties from both sides met at the Texas Hotel in Fort Worth for the expressed purpose of resolving the differences. When they could not reach an agreement, eleven co-ops chartered the Brazos River Generating and Transmission Cooperative in February 1941, and applied to REA for a loan to build a transmission line from the recently finished Possum Kingdom hydrodam, a PWA project. Charles Falkenwald, a friend and ally of Craig, had inspected the proposal, and on March 8, 1941, Craig, acting in Slattery's absence, approved it.[50]

Upon his return Slattery delayed the project for three months; the Brazos River Cooperative officers seethed with anger. Slattery also put himself at cross-purposes with Texas Congressman W. R. Poage in whose district the generating cooperative was located. Officers of the Brazos River Co-op had suspected, too, that Elliot Roosevelt had sabotaged the project because of his friendly relations with the Texas Power and Light Company, the competitor for the power generated at the Possum Kingdom dam and principal opponent of the new co-op. Congressman Poage had maintained pressure on REA to get the project started; Speaker of the House Sam Rayburn did the same, and Slattery found himself fighting the powerful Texas delegation. Craig had used the opportunity to advance his own standing while Slattery floundered in search of an escape route. The transmission line was built, but Slattery lost his standing with the Texans.[51]

When the Slattery-Ellis dispute broke out a year later, the Texas incident had further ramifications. The REA cooperatives in the state had formed the Texas Power Reserve Association with the purpose to counteract the influence of Texas utility companies in matters pertaining to rural electrification. Several officers of the new organization were charter members of the Brazos cooperative, and Ellis was able to exploit their bitterness toward Slattery for NRECA's purposes. The association sent a resolution to Roosevelt, Wickard and Rayburn denouncing Slattery and charging that he was "mentally ill, incompetent and ruining the program."[52]

The association thereby added another dimension to the feud; the cooperatives heretofore had not involved themselves, but were now beginning to demand Slattery's removal. The Texas incident also drew the question over the generation of energy into sharper focus. Since it urged REA development of power generation, NRECA was portrayed as the agent able to fulfill the original goal of serving every farm and rural home with electricity. Because Slattery opposed construction of generating co-ops, he was regarded as a stumbling bureaucrat blocking progress.

Wickard meanwhile persisted in his efforts to remove Slattery, and the latter was caught in a withering crossfire. The cabinet officer had no ideological interest, but thought Slattery's health required his dismissal. Wickard asked for his resignation, and Slattery, although surprised at the request, refused on the grounds that since NRECA charges were pending against him, he should remain at his post until they were decided. Foiled again, Wickard appointed a new deputy administrator, William J. Neal, without consulting Slattery. Neal had been superintendent of a poorly-managed cooperative in New Hampshire, had run for governor as a Democrat and lost, but was well-liked at the White House. Wickard gave Neal blanket power, and he promptly took over the functions of the REA administrator. "Slattery has been left in a kind of solitary grandeur," reported one newspaper, "fully titled, apparently legally empowered to run all of REA, yet impotent to act."[53]

The beleagured president hoped to avoid entanglement, but pleas for Slattery's removal came from all directions. Amid the confusion, one complaint about the REA chief was consistent: he could not remember his commitments and contradicted himself from day to day. When Slattery told the *Kansas City Times* that his troubles were caused by "fourth termers," the "palace guard" seeking to manipulate REA in order to promote another term for Roosevelt, the president had enough. In mid-1943 he instructed White House assistant Jonathan Daniels to secure Slattery's resignation.[54]

The first of several meetings between Slattery and Daniels occurred on July 16, 1943, when the latter asked for the administrator's resignation on grounds of poor health. To White House surprise, Slattery refused, replying that the president "ought not to throw me to the wolves." In their conversation Slattery was particularly resentful of Wickard, blaming him more than anyone else for the situation. Daniels concluded that Slattery was "off his rocker," but considering all he had been through, "it was no wonder if he was."[55] Slattery was also surprised since he did not expect such a request; he thought Daniels was "a friend," but proved to be "a two-by-four assistant at the White House."[56]

Slattery had reason to resist even though he now had to count the president among his foes. The Senate had moved to make a full-scale investigation of REA, and the investigating committee was unfriendly to Roosevelt. Judson King predicted "the investigation will be a bummer," and Daniels warned Roosevelt to hope that "the investigation of REA would turn out to be a flop and a fiasco."[57]

Hearings opened December 13, 1943, and "Cotton Ed" Smith of South Carolina, whom Roosevelt had tried to unseat in the purge of 1938, sat as chairman. It was a foregone conclusion that his purpose was to embarrass Roosevelt. Committee member Senator Guy M. Gillette of Iowa had also been a target of the 1938 purge; vengeance, not impartial judgment, was also his intention. Republican George Aiken of Vermont was fond of REA, but no admirer of the New Deal; the same description applied to Minnesota Republican Henrik Shipstead. Mississippi Senator Theodore Bilbo was the president's only friend on the committee, but he rarely came to meetings. Counsel for the committee was Carroll Beedy, a Republican and former member of the House. His assistant was Louis R. Glavis, long-time friend of Slattery, who had recently been a member of the REA staff. Prejudice for Slattery was apparent.[58]

For six months, the committee went through the history of REA since 1939 when Slattery took over. Testimony was taken, and letters, memoranda and reports, textual materials of all sorts were filed as evidence, until the official transcript came to more than 2,000 pages. Each major participant testified; the hearing was an occasion for witnesses to repeat their accusations. Slattery accused Wickard of infringing upon his prerogatives and reversing his decisions, and of using REA for political purposes. He accused Ellis of attacking him for his refusal to approve the NRECA insurance plan, and of seeking to defame his character when all else failed. Wickard replied that Slattery was an inept, blundering administrator, that as far back as 1941 Roosevelt had approved his suggestion to remove Slattery. Ellis denied any intent to secure control of REA for political purposes, pointing out that he had offered to remove any objections to the proposed insurance plan. He urged Congress to return the agency to independent status. As part of its annual policy statement, the Farm Bureau also urged that REA be kept out of politics.[59]

Daniels was subpoenaed, but on grounds of executive privilege refused to disclose the full nature of the president's part in the request for Slattery's resignation. Suddenly the hearing made front-page headlines because the senators, Daniels observed, thought "they had a good chance to hit the Administration."[60] The committee unanimously proposed that Daniels be cited for contempt of the Senate, but the White House assistant was not "greatly disturbed by the prospect of jail." Roosevelt wanted to avoid a confrontation for the Senate was ripe for rebellion as indicated by Kentucky Senator Alben Barkley's temporary resignation as majority leader in protest of a presidential veto. Roosevelt instructed Daniels to reveal his part in the scenario, but since Daniels had nothing startling to reveal, his testimony was anticlimactic. The tension evaporated, and Daniels recalled that he "even had old 'Cotton Ed' Smith smiling."[61]

After Daniels's appearance, witnesses continued to lash out bitterly at one another, but the proceedings now had no fanfare and ended in

May 1944. The committee's report exonerated Slattery, blaming the disintegration and demoralization of REA on Wickard. Charges of influence-peddling on behalf of the copper interests were accepted as true, and the evidence, regardless of the bias of the committee, seemed to verify that conclusion. But the committee implied that USDA, not Slattery, was to blame for the copper incident because Secretary Wickard has usurped his authority and responsibility, leaving him unable to act. The intense attack on Slattery, the report continued, originated with NRECA because he refused to approve the insurance plan, a decision, the committee pointed out, the USDA solicitor's office has reinforced. While commending Slattery's refusal to surrender, the report conspicuously omitted references to the charges against him: poor leadership, absenteeism, indecision, contradiction and favoritism of certain employees. It did not consider the question of power generation. Wickard and Ellis were blamed for the whole incident. A more complete report, however, would have brought out the importance of all the charges and issues involved.[62]

If the committee absolved Slattery of the charges against him, he had won a pyrrhic victory at a point of no strategic significance. When Daniels had asked Slattery to quit, he was, in effect, showing the president's approval of NRECA. The known bias of the senators removed the sting of the indictments against Slattery's foes. This last point was obvious in that Slattery received no help as Wickard continued to hamstring him. A flamboyant hunger strike by Chester Lake, one of the alleged "pensioners" and long-time REA employee whom Wickard fired, failed to generate new enthusiasm for Slattery. The public was more interested in the progress of the war for the hearing had closed shortly before the allied invasion of Normandy. Slattery had become totally isolated, out of favor with the president and most REA supporters. His desperation was evident when he supported Thomas Dewey in the 1944 campaign, and only after Roosevelt's re-election, did he resign on November 25, 1944.[63]

Slattery's isolation was evident with the passage in 1944 of the Pace Act which set the REA rate of interest at 2 percent, a move which in later years was responsible for the charge that REA was subsidized. Discussion of the need for subsidy went as far back as the origins of the agency, and the idea received new attention during World War II since REA was expected after the war to serve every farm and rural home, however poor or remote. NRECA lobbied for such legislation in 1943, but Congress refused to pass it. At the time, Rayburn had asked Slattery for his thoughts on subsidization. "The REA program," he replied, "does not need subsidy now or in the predictable future."[64] Once again it was evident that Slattery and NRECA disagreed over the fundamental role of REA.

In 1944 the subcommittee of the House Agriculture Committee reviewed proposed REA legislation. Officers of several agencies and NRECA testified, but Slattery was absent because Wickard insisted on handling REA affairs with Congress. Representatives of the electrical industry did not appear at the hearings. The subcommittee recommended removal of the ten-year time limitation on REA and called for the establishment of a 2 percent interest rate on REA loans to cooperatives. Subcommittee chairman Stephen Pace of Georgia thought the proposed interest rate justifiable since the Treasury Department furnished funds at comparable rates to RFC, Federal Farm Mortgage Corporation and other federal agencies. Texas Congressman Poage expressed the most common view of the proponents when he said agriculture deserved the same treatment as industry. The measure breezed through the House.[65]

In the Senate, the measure sped through the Senate Agriculture and Forestry Committee which, ironically, was involved through its subcommittee with the investigation of REA. "Cotton Ed" Smith chaired both committees; he defended Slattery in the subcommittee to avenge Roosevelt's attempted purge of him in 1938, but the senator disagreed with the REA administrator over subsidization because he thought poorer homes and farms should have service. Almost as a matter of routine the provision for the 2 percent interest rate passed Congress and became law on September 21, 1944.[66]

Liberalization of the credit terms encouraged the establishment of new cooperatives and allowed existing ones to extend their territory. It also meant that families living at the bottom of the agricultural ladder, the sharecroppers, renters and laborers, were more easily within reach of the comfort-giving current. It was for this class that the fight for power generation and subsidization had been fought, and Slattery's failure to understand this point cost him his job.

Given its myriad of problems, REA should be praised for its achievements and not criticized for any failures. At the end of 1944 when Slattery stepped down, the rate of connecting new farms was approximately 50 percent of the prewar average. In 1944 farms with REA service numbered

1,216,798, almost treble the number in 1939 when Slattery took over. It was not true that he was ruining the program even if he disagreed over the future role.[67]

REA success in the midst of the war and internal squabbling was due to the recognition that electrical machinery was one of the best substitutes for manpower. In convincing the wartime priority agencies of that fact, Slattery, Wickard and Ellis worked not as a team, but as individuals, each carrying the same message: "release us from your restrictions." However much they disagreed among themselves, their effect was the same in so far as lifting the restraints. Credit for the efficient use of the reduced appropriations must be given to the dedicated REA staff. Self-seeking promoters contributed to REA's problems, and without them the program would have probably been even more efficient. Contributing to the success of the agency were the men and women in the field, from the co-op superintendents to the journeymen electricians, for all were enthusiastic over the program. Despite the upsetting conditions, the agency still moved forward, and if Slattery was weak and indecisive, he was lucky that the great bulk of his employees were only interested in their jobs.

Until his death ten years later, Slattery blamed his woes on political hopefuls and ambitious bureaucrats who sought to make the agency a base for empire building. Allegedly they conspired to manipulate the 4,000,000 members of REA for political purposes, or to use the program for monetary gain. He attributed the former to Wickard; to Ellis the latter. Such was not the case; Wickard served as REA chief when Harry Truman was president and never attempted to exploit the votes of REA recipients. So widespread was the area of REA operations and so diverse the rural population of the United States, that no single person or organization could take advantage of the REA program for political benefit.

Charges against NRECA never proved true. The association found satisfactory insurance coverage for the cooperatives with Employers Mutual of Wausau, Wisconsin, and dropped its own plan. No scandals or profit-making schemes were associated with Ellis personally, or with NRECA as an organization. At times belligerent, due to Ellis's style, the association became the respected lobby for public electrification.

The Slattery incident was a turning point regarding REA. New interpretations of the agency's role came at an inopportune time in so far as national harmony during the war was concerned. Slattery's service in the agency also came at an inopportune time. He belonged to the old guard, "highly deserving as an oldtime progressive" as Lilienthal wrote, but his health was failing, and he was unable to cope with the pressing situation in which he was thrust. His personal difficulty was sad, demanding sympathy, but the relentless succession of events gave none, and he became a tragic figure.

If the incident was a particularly intriguing link in the chain of events leading to complete electrification, it was not the final link. Despite REA's move into power generation, use of electricity was still hampered by the shortage of energy, particularly in the South. Hoping to alleviate the shortage, proponents of electrification were already trying to develop hydroelectric sites, but they became embroiled in a pitched battle with the electrical industry.

9
THE POSTWAR FIGHT
FOR ENERGY
IN THE SOUTH

During the course of the war, shortages of energy had occurred in rural areas causing severe reductions in usage and even occasional outages. Shortages were most acute in the South where energy development had not kept pace with demand. A significant increase in demand was expected at the war's end when REA resumed full-scale its construction program, a development made even more probable, of course, by the liberal credit terms set forth in the Pace Act. Potential water power sites were numerous in the South and offered excellent opportunities to supply cooperatives with adequate energy. But the utility companies opposed further development of hydroelectric power, and when a drive started to utilize these sites, a stiff contest occurred. The fight centered on the establishment of the Southwestern Power Administration (SPA) and to a lesser extent on the Southeastern Power Administration (SEPA). These two agencies, particularly the former, were illustrative of the national fight for supplies of electricity in the postwar period and must be considered part of the push for complete rural electrification in the United States.

Initial steps toward the creation of the SPA were taken as far back as the 1920s when local citizen groups consisting of town and country merchants, chambers of commerce, farmers and mayors started a campaign to build flood control reservoirs with hydroelectric dams on the Red River at Denison, Texas, and on the White River at Norfork, Arkansas. Flood control was important to the sponsors of the projects, but their main purpose was to generate power for rural electrification.

As was the case in many states, fewer than 3 percent of the farms in the vicinity of the proposed reservoirs had service, and rates charged by utility companies were too high to encourage farmers to electrify their premises. The companies had no intention of serving them and proponents of the dams regarded public power as the solution. A spokesman for the Denison group told the federal House Committee on Flood Control in 1930: "The greatest benefit that comes out of impounded water is . . . the 'juice' . . . you can light up that whole country and turn every barn into a factory by giving the farmers the power . . . we will milk the cows, run the refrigerators, rock the cradles, fry eggs, and bake the cakes with electricity."[1]

Congressman Rayburn, whose district included the site of the proposed dam at Denison, agreed the only solution was public development and operation of the water resources in the area. After a long and persistent campaign, conducted while he also sponsored the REA bill in 1936, Rayburn won congressional approval of the dam in 1938. It was built during World War II and went into operation in June 1944.[2]

Clyde Ellis was the principal proponent of the Norfork project and wanted to model it after TVA. It was with this project that he won his congressional seat in 1938 and it was the chief project of his political career. He arrived in Washington while Rayburn was maneuvering for approval of the Denison dam.[3]

Among Ellis's first contacts was Rayburn, at that time House majority leader, and the dependence of the freshman representative on the veteran legislator was an example of the latter's interest in rural electrification outside his own state. Ellis soon had an opportunity to see Rayburn in action.

It would not have been difficult for either to secure approval of an earthen dam, but to win approval of a power plant was another matter. Rayburn had scheduled a meeting with General Julian L. Schley, chief of the Army Corps of Engineers, to discuss the matter, and he invited Ellis to attend. As Ellis listened, Rayburn went to the point: "General," he said, "this boy . . . and I want power put in our dams and we mean to get it." General Schley was far from encouraging. "The power companies are opposed," he replied, "and they'll fight it." Rayburn was not discouraged. "We'll take care of that," he replied. "You just get your house in order, and let's go for this authorization."[4]

Within a short time the Corps had drawn up plans for hydrogenerators in both structures. Ellis won approval of the Norfork dam in 1940. Generation of electricity for test purposes began in June 1944, shortly after

Ellis lost his Senate race and joined NRECA.[5] Such quick success for a newly-arrived member of Congress was more than unusual—it was rare, demonstrating his qualities as a legislator. The incident with General Schley, however, showed how Ellis, despite his own ability as a promoter, benefited in no small way from Rayburn's help.

America's entry into World War II compounded, of course, the shortage of energy in the Southwest as elsewhere. It was this problem that had prompted NRECA to urge Slattery to move REA into power generation, and the fight over the Brazos River Generating Cooperative was part of the same problem. The structures at Denison and Norfork were not part of the Slattery incident because Rayburn and Ellis kept them out of the REA internecine war, but they were directed toward alleviating the shortage. Slattery's lack of contribution in this drive for more energy was further evidence of his isolation.[6]

As the building of the dams progressed during the war, they acquired greater significance because of their relation to the future development of public electrification in the region. By 1943, shortly before either dam was completed, Congress had authorized the construction of eight additional hydroelectric plants in Texas, Arkansas and Oklahoma. More were expected. Rayburn and Ellis were convinced that the original projects should be integrated with the newly-authorized dams in order to form a large grid system, which, they reasoned, would provide better service for the rural market.[7]

Utility companies in the area, however, opposed the development of a public grid, for the same general reasons as did most companies faced with competition. They resented intrusion upon a market they had monopolized and they expected that lower rural rates would go into effect with the use of less expensive water power. Conceivably they might have won jurisdiction over the dams, for as late as 1943, when the plants at Denison and Norfork were nearly completed, Congress had not designated any authority to operate them permanently after the war. The utilities wanted to acquire jurisdiction over the two original dams since the final decision about them was expected to apply to the additional structures already approved by Congress.[8]

Private interests seemed likely to win control because as each dam was finished, it was placed temporarily under the Federal Works Administration (FWA) which was responsible for supplying defense industries with electrical power. Earlier the Grand River dam in Oklahoma, which also originated as a local campaign to increase energy, had been placed under

FWA; FWA controlled the Norfork structure, and the Denison dam seemed destined for the same fate. Rural electric spokesmen thought FWA was sympathetic to the electrical industry since administrator Philip B. Fleming, a career Army Corps engineer, favored the private interests. Rayburn and Ellis were worried that projects placed under FWA for defense purposes would remain there permanently. When General Fleming proposed in May 1943 that President Roosevelt put the Norfork and Denison dams under FWA permanently, their fears seemed well-founded.[9]

The struggle over the dams in the Southwest was part of an even larger fight within the Roosevelt administration over control of government-developed energy resources. The final decision about the operations at Denison and Norfork rested, therefore, on the outcome of that contest. Secretary of the Interior Ickes wanted federal reservoir projects placed permanently under his jurisdiction, arguing that only his department could keep them out of private hands. Ickes also wanted authority to set rates, a prerogative of the Federal Power Commission (FPC). Leland Olds, chairman of FPC, suspected that Ickes was trying to establish himself as an autocratic ruler of public power. From Olds's standpoint, Ickes's ambition threatened FPC. Details of this struggle between bureaucrats of both agencies for jurisdiction over power projects are little known and inconclusive, but the evidence suggests that more than anything else it was a matter of departmental jealousy, for Olds, too, favored public control of water resources. In any case, Olds backed Fleming's move, perhaps to block or delay Ickes's seizure of the dams. This difference of opinion in high councils of state, regardless of the cause, had the effect of postponing a decision regarding the control of electricity for rural use in the Southwest.[10]

Because of the tardiness in reaching a decision in Washington and because he feared loss of the dams to FWA, Rayburn took steps to ensure the rural use of the power. To protect the structures against future manipulation, he wanted them permanently placed in the Department of the Interior, which he considered the logical agency to administer the use of natural resources. Also, since Ickes was such a public power zealot, Rayburn saw him as a valuable ally. The Department of the Interior would more likely be amenable to pressure from Congress than would FPC or FWA in future disputes over control of the energy.[11]

Private interests were not expected to make their bid for control of the dams until the war ended, and their move precipitated a swift struggle over the Southwest's energy resources. Now Speaker of the House, Rayburn retained as counsel in power affairs, Alvin J. Wirtz of Austin, a former undersecretary of the Interior. Wirtz drew up a presidential order placing the dams in the Interior Department, which Rayburn discussed with Roosevelt in July 1943.[12]

After his conference with the president, Rayburn kept up the pressure. From his home in Bonham he wired the president that:

> For reasons sufficient to me and which I can explain to you when I
> see you I do not want these dams under the Federal Works Administra-
> tion as I understand had been suggested. If you cannot put this under
> the Department of Interior now, I will be glad to discuss it with you be-
> fore you put it under any other authority. [13]

The Speaker used political muscle in other ways. Three days after contacting the president, he wired James F. Byrnes, at that time head of the Office of War Mobilization. Explaining the situation, Rayburn requested that Byrnes, "See about this and if the President cannot sign the executive order placing it under Interior, please ask him to hold it until I return."[14]

A few days later, Roosevelt wired Rayburn that he had just placed the dams at Denison, Norfork and Pensacola under the Department of the Interior. The executive order authorized the Secretary of the Interior to sell and dispose of electrical energy generated at the dams to war plants and establishments, public bodies and REA cooperatives, and utility companies, in that order of preference, at such rates as were approved by FPC. Thus, when the war ended, the power from the dams in the Southwest was reserved for use by farmers ahead of utility companies.[15]

The next step in the organization and control of the water resources came with the creation of SPA. Acting on orders from the president, Ickes created the agency on August 31, 1943, and appointed Douglas G. Wright of the power division of the Department of the Interior to administer it.[16]

The location of the dams in the Department of the Interior and creation of a special agency to administer the use of the power generated at them was the first and most important clash in the fight to guarantee farmers in the area an adequate supply of reasonably priced electricity. This accomplishment was reached without the dams becoming involved in the REA contest over power generation. It was due to Rayburn's political clout that he could maneuver them as he wished, especially since Roosevelt had ignored the urgings of cabinet members and high level bureaucrats to put the dams on a permanent footing.

Rayburn's influence with Roosevelt was the product of their working together for over ten years. When the Texan was chairman of the House Committee on Interstate and Foreign Commerce during the 1930s, he had worked closely with the president in securing passage of legislation as significant as the Securities Exchange Act and the Utility Holding Company Act. Roosevelt knew, too, that Rayburn had a crucial role in his nomination by the Democratic party in 1932. In 1940 the Speaker had cast the deciding vote to institute the military draft, a measure Roosevelt had regarded as essential for preparedness. During the war, the president was pleased to have Rayburn as Speaker of the House.[17]

But Roosevelt also knew that Rayburn represented the rural electric interests, for when the latter justified the dams of SPA he spoke for thousands of individual farmers pleading for more electricity, plus REA co-op managers and leaders of agricultural organizations. Typifying the grassroots support of SPA was a letter from one constituent. "You got this project for us, and I sincerely hope you will see to it that only the government controls the power from the dams."[18] "I was raised on a farm," wrote a nonconstituent, "and when it comes to appropriating money for the development of electricity made from water . . . you can't spend too fast to suit me."[19] From the Grange in Fowler, California, he received this commendation: "We most heartily endorse your action . . . so that the millions of tillers of the soil may get a fair break in the use of these great projects."[20]

The dams were safely in the hands of the public interests, however, only as long as Rayburn remained on good terms with the executive branch since the presidential order was only a temporary measure. Until Congress placed the structures under the Interior Department permanently, there was always a chance that they might be transferred to an agency unfriendly to public power. This dilemma was resolved when Congress passed the Flood Control Act of 1944, a comprehensive measure that gave the Secretary of the Interior jurisdiction over federal reservoir projects that were supplying defense industries with electricity. It also authorized nationwide construction of more dams after the war, though not all in the Southwest. The total authorization was one billion dollars.[21]

Congress passed the measure in order to establish permanent jurisdiction over the energy generated at the plants either in operation or to be constructed in the future. No one objected to the defense use of the power during the war, but in the Southwest and elsewhere postwar use of energy was another matter. The Flood Control Act of 1944 gave public bodies, which included REA co-ops, first use of the power. Privately-owned companies were not excluded, but they were eligible to draw upon this supply of electricity only after the needs of the public bodies were met. Ickes had written the provisions concerning the prerogative rights to the energy and patterned them after the provisions of SPA.[22]

When SPA was created in 1943, administrator Wright prepared a comprehensive plan to make power available to cooperatives as soon as the war ended. He first had to devise a way to overcome a unique problem to power agencies in the Southwest—the lack of a sufficient year-round supply of water. During the months when the water level was low, SPA needed to acquire power from other sources in order to guarantee a steady supply of energy, or "firm power," to its customers. When the water level was high, SPA had more than enough energy and needed to dispose of the surplus, or "dump power."[23]

The agency had the option of building steam-generating plants to supplement the hydroelectric dams, but it sought to overcome the difficulty by exchanging dump power for firm power from the utilities. SPA, in other words, would give its surplus energy to the utilities if they would supplement the hydroelectric plants during the dry months. This proposed system of integrated facilities joined the water-generated power of the dams with the fuel-generated power of the private plants, a method recognized as an efficient and advantageous utilization of resources as long as the private and public interests cooperated.[24]

The decision to integrate was not made to appease the utility companies but was a manifestation of Rayburn's ideology. It conformed to his home-spun middle-of-the-road political philosophy. Both parties should work together, he reasoned, and public power projects should operate alongside investor-owned plants. "I don't belong to either group," he would say; the cooperatives and utilities should resolve their differences.[25] He had expressed the same concept during the course of the House debate over the REA bill in 1936.

Ellis of NRECA disagreed and wanted Congress to create a "little TVA" in the Southwest. As a member of Congress he had introduced a bill for an Arkansas Valley Authority, and he supported proposals for a gigantic Missouri Valley Authority. Chances for either were slim, but he hoped to realize his dream through SPA. He envisioned an enlarged and more encompassing agency reconstructing the area along the same lines as TVA. In this respect, Ellis and Rayburn were ideologically opposed to one another, demonstrating a difference of economic philosophy and opinion among rural electric leaders. Congressman Rankin belonged to the more radical group; he introduced a measure in 1945 providing for the creation of eight valley authorities patterned after TVA, including an Arkansas Valley Authority in which both the White and Red Rivers were tributaries.[26]

As part of the contingent lobbying for the "little TVA's," Ellis and Rankin saw river basin development as an opportunity to revive rural America. Migration into the cities would end only by making farming both economically and culturally rewarding. Massive reconstruction programs, therefore, were needed to combat lowering agricultural prices, rising operating costs, erosion of soil and the general cultural decay of rural areas.

For Rayburn, time, circumstance and postwar prosperity removed any need for rehabilitation programs designed for conditions of a generation earlier. Federal assistance agencies already in operation had ended much of the deprivation and were expected to continue doing so. Farmers only needed power, and SPA administrator Wright agreed: "the policy to be followed," he told Rayburn, "should be governed by your judgment."[27] President Truman, a close friend of the Speaker, showed no enthusiasm for the "little TVA" and supported SPA as envisioned by Rayburn.

Proposals for the valley authorities characteristically called for rather large bureaucracies independent of the usual civil service regulations. That, plus the notion such authorities should plan so as to effect the "economic, social and cultural values" in their area of operations not only went beyond the matter of rural electrification but also raised the question of federally imposed "superstates." In the midst of the nationwide discussion over river basin development, Rayburn proceeded to fight for piecemeal expansion of SPA solely as an energy supplying agency. Ellis was overwhelmed by Rayburn's clout, but he still worked hard on behalf of SPA, using the resources of NRECA on its behalf.[28]

Proposals for a "little TVA" naturally stiffened the resistance of the power companies and when Wright presented a plan to integrate SPA with their facilities, he met tough and well-organized opposition. Utility executives made "violent attacks on the radio; in newspapers and magazines; in the mail; at employee and union meetings; in Congress; and in the courts."[29] The most dramatic test came in 1946 when power companies tried to defeat the next year's SPA appropriation. Wright had requested $23,000,000 from the House Committee on Appropriations in order to combine the facilities at Denison and Norfork. Company spokesmen charged that SPA duplicated resources and infringed upon the prerogatives of private enterprise. Congressmen were swamped with propaganda to persuade them not to vote for the appropriation. It was alleged that power companies sent nylon stockings to the wives of congressmen, items in short supply in 1946.[30]

In its final report, the committee trimmed the budget to $3,198,000, an amount too small to enable SPA to start its program. At this point the agency appeared doomed since committee reports usually had the effect of a life or death sentence. Former Secretary of the Interior Ickes described the committee's decision as "The most wanton disregard of public interest since the famous 'public be damned' statement by Commodore Vanderbilt."[31]

When the committee reported on the measure, Rayburn took an unusual step. Rarely did he leave his chair as Speaker to participate in House discussions, but when the SPA appropriation measure reached the floor he laid down his gavel and took his place as a regular member. He offered an amendment to the appropriation bill to raise the funds allotted to SPA to $7,500,000, enough to permit the agency to start a limited program of integration. "I deeply regret," he said, "that in the interest of . . . the Federal Government, I feel called upon to offer an amendment . . . not to parallel anybody's lines, not to put anybody out of business, but simply to tie this Government property together . . . to electrify our homes and serve our industries." The amendment carried.[32]

The connection between SPA and the broader purpose of rural electrification was obvious. On an earlier occasion Rayburn had left the Speaker's chair during the House consideration of the REA budget for 1947. It had been increased substantially to $250,000,000, and the electrical industry vigorously objected. "This town has been seething with utility lobbyists," Rayburn stated, and "they are now before a subcommittee . . . where we have asked for a little appropriation to carry on the business of the SPA. They are in there trying to kill it off." Referring to the contest over the

Public Utility Act, he added: "I had a brush with these people in 1935. If they are spoiling for another fight with me . . . they can get it."[33] Rankin of Mississippi and other public electric proponents had fought for the larger REA appropriation, and after Rayburn's remarks the House passed it.

It was clear in other respects that rural electrification was the focal point of the clash over the SPA appropriation. Although friends of public power everywhere supported SPA, the strongest spokesmen on its behalf were representatives of REA cooperatives in Texas, Oklahoma and Arkansas, who pleaded with members of Congress to support the appropriation so that farmers could enjoy the benefit of inexpensive electricity. The National Farmers Union presented petitions from its members in Texas and Oklahoma urging approval of Wright's budget. At the same time, the organization also endorsed the bill for a Missouri Valley Authority. Ellis regularly testified, of course, for SPA before congressional appropriation committees, and state associations such as the Texas Power Reserve endorsed the agency. So numerous were the spokesmen that the committees had to limit the amount of time for each to speak.[34]

This struggle over the appropriation was followed by other attempts to cripple SPA. Typically, opponents of the agency attacked it as a socialist concept endangering free enterprise. Frank M. Wilkes, president of Southwestern Gas and Electric Company, writing in *Electrical South,* implied that SPA was another example of "Fabian tactics that have succeeded in undermining a considerable portion of our free enterprise system." Among the friends of the utilities were Oklahoma Senator E. H. Moore and state representative Russell Smith. Moore even proposed that SPA be totally liquidated and its physical plant sold to the power companies.[35]

Utility companies were quite forceful in fighting the expansion of public electrification, and only because they faced Rayburn were they not more successful. His friendship with Truman was beneficial; ordinarily he either discussed each situation concerning SPA with him or sent a formal letter, or both. Truman's feelings on the subject were, as usual, made clear in his statement at the dedication of the Bull Shoals dam in Missouri, part of the SPA system: "The power companies, who said the dams were socialism, are not passing power along to their rural customers. Power ought to go to the farmers, and as long as I have anything to do with it, that's where will the power will go."[36]

While battles raged over appropriations in Congress, administrator Wright negotiated with the utilities to exchange energy. They persistently refused, hoping to see the complete liquidation of the program, or at least to force the agency to sell the power on their conditions. The beginning of a break in the attitude of the private suppliers came in 1947, however, when the Texas Power and Light Company (TP&L) signed the first contract; Public Service Company of Oklahoma soon followed. The remaining companies did not suddenly end their feud with SPA, but the example set by TP&L, the largest company, had a discernible softening effect on their attitude. As Rayburn told Truman, "This contract will be used as a pattern for negotiating with the other companies." The TP&L contract, furthermore, furnished a large enough exchange of energy to absorb the generating capacity of the newly-built dams, enabling SPA to continue its plans. The rest of the companies were slow to make agreements, but as they saw TP&L enjoying the benefits of the integrated system, and as they realized that SPA's only interest was the rural market, they signed contracts. By 1952 opposition was rapidly disappearing, and the president of a company which had led the fight against the SPA earlier, admitted, "If I appear before the Congressional Committee on Appropriations for fiscal year 1953, it will be to support the entire program."[37] Ellis, ideologically opposed to joint agreements, tried to prevent the TP&L contract, bringing a rebuke from Rayburn. Ellis's attitude softened as he realized he was defeated and when he saw the arrangements working satisfactorily.

Throughout the pitched battle in the Southwest, Ellis fought on behalf of appropriations for the Bonneville Power Administration (BPA); Rayburn and other congressional friends of REA joined him for despite their ideological differences, they shared a comradery in their belief of public electrification. Events in the Northwest, however, never engendered the bitterness found in the Southwest, nor was rural electrification so much the focal point. To be sure, REA co-ops received power from BPA, but the earlier federal commitment to public power on such scale as found in the Northwest had ended the question there.[38]

As new dams in the Southwest were completed they were placed in SPA, and the agency grew, integrating the generating plants and transmission lines of public and private interests. No steps were taken to make it into a "little TVA." By 1969 SPA had twenty-three dams, with a total generating capacity of 2,131,000 kilowatt-hours, serving the six-state area of Texas, Oklahoma, Arkansas, Missouri, Louisiana and Kansas.[39]

The impact of SPA demonstrates that the original intention of the founders has been accomplished. Two achievements are clearly visible: (1) an ample supply of electricity for farmers, and (2) rates low enough to be within the reach of the average farmer. By 1968 REA cooperatives received 53 percent of all SPA electricity. The rest of the power was distributed among municipal districts and some private interests.[40]

The cost of energy supplied by SPA has steadily declined. In 1941 the average wholesale rate for rural cooperatives in Texas and Oklahoma was 1.03 cents per kilowatt-hour. By 1945 this average had dropped to 0.80 cents, and by 1950 the average rate was further reduced to 0.57 cents. Such rates were comparable to those of the Tennessee Valley Authority and lower than those where a supply of public power was unavailable.[41]

A similar but less dramatic story took place in the southeastern United States with the creation of the Southeastern Power Administration (SEPA) in 1950. That year two federal dams were in the final stages of construction, the Dale Hollow project in Tennessee and Allatoona project in Georgia. Both originated with congressional authorization as in the case of the Denison dam in Texas. Construction had begun on another seven dams located in South Carolina, Kentucky, Tennessee, Virginia and Florida, and another fifteen projects already had congressional authorization. An agency was needed to administer the operation and sale of power from these projects once they were finished, and the SPA conveniently served as a model. Following the precedent set in the Southwest, the Secretary of the Interior created SEPA by executive order in a rather matter-of-fact fashion, basing his authority on the Flood Control Act of 1944. Preference in the sale of power went to the cooperatives and public bodies.[42]

The new agency controlled the operation and sale of power in West Virginia, Virginia, South Carolina, North Carolina, Alabama, Mississippi, Kentucky, Georgia, Tennessee and Florida. In those instances in which SEPA jurisdiction overlapped TVA, the former sold its power to the latter which assumed responsibility for distribution. Although SEPA was established without fanfare or theatrics, it was, nonetheless, identified with rural electrification since NRECA and the Southeastern Power Committee

(SPC) were its principal proponents. The latter originated in 1942 when the REA cooperatives in eastern North Carolina established a "power pool" to supply defense plants and military installations. After the war, SPC focused attention on farm energy needs, so that by 1951 it boasted a membership of 212 cooperatives in ten states. For the most part SPC was influential only in the annual dispute with the electric companies operating in the area over congressional appropriations for the agency.[43]

Battles over appropriations were intense until about 1955 when the utilities resolved that public power and transmission lines were firmly established in the Southeast. Like its sister agency in the Southwest, SEPA developed an integrated grid with private sources of energy. Rates steadily declined until they reached levels comparable to those of TVA and SPA.[44]

Establishment of these two agencies was necessary for the complete electrification of farms in fifteen states where energy supplies were lowest. Ellis had referred to the areas as "power deserts." By providing a competitive source of energy, they also reduced the rural rates of the power companies. The significance of these agencies, particularly SPA, extended beyond their area of operations because the time and circumstance of their origin coincided with the post-World War II debate over public electrification. Just as the architects of SPA laid the foundation stones of their agency, beginning in 1943 and continuing for the next decade, the electrical industry hoped to regain its position as the instigator and provider of electric service. Considerable attention was attached to SPA, making it a weathervane of the political winds. Had Rayburn sought to establish a broader agency as envisioned by Ellis, he would undoubtedly have lost. Despite their differences they worked as a team, and the establishment of SPA in view of the opposition to valley authorities was a testament to their cooperation.

In the development of rural electrification, political muscle was the key factor of Rayburn's contribution. His political wisdom and strength, ranging from support at home to a close relationship with two presidents, were the base of his success. His constituents gave him an intense and fierce loyalty, and he drew upon their devotion to show popular enthusiasm for the projects. His easy access to the White House came from his long commitment to the Democratic party and his key support for the occupants of the Oval Office.

10
AMERICA ACHIEVES RURAL ELECTRIFICATION

There are many bright spots on the postwar market horizon, but probably none hold forth more promise than rural electrification, which is mutually beneficial to both rural populations and private industry. By relieving many of the physical burdens that have been the heritage of rural America for generations and by increasing agricultural production and income, rural electrification can bring about a higher rural standard of living.[1]

This opening statement of a special postwar market forecast in 1944 by *Country Gentleman* expressed the widely-held opinion that every farm and rural home should have electricity. As of January 1, 1944, 55 percent of America's farms were still without service, but already there were indications that an intense REA building program would be started at the war's end to complete the task begun a decade earlier.

As far back as October 1939, former REA administrator Cooke had suggested that a long-range planning program commence after the war. President Roosevelt had expressed his opinion that such a program would be useful to fight unemployment when the war was over. In December 1944, while Rayburn fought for an energy supply for southern farms, the REA Postwar Planning Committee drew up a preliminary plan for extending service to the remaining 3,500,000 unelectrified farms and homes, estimating that the total cost for such an undertaking, including purchases of wiring and appliances by farmers would reach $5,500,000,000.

"How large this construction program will be and how rapidly it will be carried out," the committee said, "will depend primarily on action of the United States Congress."[2]

Congress had already indicated its willingness to expand REA by passing the Pace Act, and it was generous with appropriations after the war. The highest annual appropriation was $470,000,000.[3] Reasons for the generosity varied, including the work of Congressman Rankin, but the popularity of REA was the most important. Farmers without service felt they had waited long enough, and their votes were likely to express their dissatisfaction. Prodding Congress into action, furthermore, was Ellis of NRECA which was already "a force that cannot be ignored either in Washington or in the country."[4] In 1950 the Farmers Union lauded Ellis for his "outstanding service to agriculture." Endorsement of REA from the National Grange and Farm Bureau was another indication of the demand for service. Not to be overlooked was the support of President Truman. The combination of these circumstances meant that REA could expect to receive hefty appropriations almost as a matter of routine.

With funding generously provided by Congress, the reduced interest rate of 2 percent on cooperatives' loans and the steadily increasing supply of energy made available by REA generating plants and agencies typified by TVA, SPA and BPA, the rate of serving farms after World War II moved at phenomenal speed. Postwar prosperity also enabled families to acquire appliances with greater ease, causing demand for electricity to jump proportionately. By 1953 a total of 2,544,000 farms were connected to REA lines. The agency had, since 1935, loaned a total of $2,788,136,191 to 983 cooperatives, 44 public power districts, 26 other public bodies and 25 electric companies.[5] Of the subscribers, 67 percent were farmers, while the remainder were rural, non-farm establishments. Only the most remote and marginal farms and homes did not have service; for all practical purposes, American agriculture was electrified.[6]

Progress was also rapid because power company opposition to REA had subsided by the early 1950s. The electrical industry's lobby was no match for that on behalf of REA. Spite line activity had ceased due to court decisions and agreements between REA and individual companies. Industrial and population growth taxed the resources of the companies and as they frantically tried to keep up, they accepted REA as a partner in the rural market.

Rural rates showed a slight decline from their pre-World War II levels which had ranged about 3 to 5 cents per kilowatt-hour; there was also some decline in REA rates in relation to those charged by utility companies. Conditions under which both parties set rates still varied widely, but the large increase in power generated by public agencies accounted for the drop in REA rates. This competition also encouraged the private suppliers to reduce rates to their recipients.[7]

TABLE 2: **Average Revenue Per Kilowatt-Hour, REA Cooperatives and Private Utilities, 1948-53**

Year	REA Co-ops (Cents)	Private Utilities (Cents)
1948	2.44	3.01
1949	2.39	2.95
1950	2.33	2.88
1951	2.23	2.81
1952	2.17	2.77
1953	2.10	2.74

SOURCE: *Moody's Public Utility Manual* (New York, 1954), p. A13; REA, *Statistical Annual Report, Rural Electric Borrowers,* REA Bulletin 1-1 (October 1973), p. 9.

The votaries of private enterprise insisted that REA owed its success to subsidization, that only with special advantages could it equip poorer farms and homes with electricity. Defenders of the agency disagreed. To the extent that the REA rate of interest was set at 2 percent, from 1945 through 1973, the program was subsidized because interest rates paid by the utilities were always higher by 1 percent or more. REA had another advantage since the loans to cooperatives represented its entire investment, whereas power companies financed a considerable portion of their investments with costlier kinds of securities. Rates of the latter could have been lower, however, if they had been issued in serial bonds repayable in monthly installments. The cooperatives were, nonetheless, subsidized through the rate of interest.[8]

Tax exemptions were another advantage enjoyed by the cooperatives, although the exemptions varied from state to state. Cooperatives never paid income taxes, but they were subject to social security, unemployment insurance and property taxes in most states. In the case of property tax, however, states frequently assessed only the earning power of the cooperatives and not the property, agreeing that since they had a lower earning power than electric companies operating in the urban market, they must not be taxed beyond their ability to pay. REA headquarters in Washington welcomed the exemptions, but never promoted them. When some cooperatives retired their total debt to REA, their taxes went up on the grounds that they were now privately incorporated bodies subject to the same taxation as utilities. Again there was no uniform rule; each state followed its own code.[9] REA also rendered services to the cooperatives at no cost, ranging from management supervision to appliance instruction for recipients. These services constituted a small portion of operating expenses for either the parent agency or local cooperatives.

Advantages were also extended to the utility companies, usually under the guise of regulation. The theory of setting rates so as to allow a "fair and reasonable" return on the companies' investment, normally 6 percent, seldom worked out in practice. REA defenders pointed out that the rates made allowances for taxes, but most utilities also collected "overcharges," thereby producing more than the reasonable return on investment authorized by regulation. Not only were overcharges commonplace, it was argued, but the companies kept two sets of accounts, one for making rates and another for tax purposes, the latter being designed for a "liberalized depreciaiton" which achieves the result of lower income taxes. Such tax benefits, NRECA insisted, should be considered the same as interest-free loans from the federal government. Investment tax credits, designed to stimulate plant expansion, constituted a federal subsidy. Controversy and confusion have thus characterized the debate over the comparative advantages enjoyed by REA co-ops and the electrical industry. In the final analysis the cooperatives were and are still considerably public in character, but their advantages have not been as extensive as alleged.[10]

Regardless of special advantages, the record of REA as a lending agency has been and continues to be outstanding. With less than 1 percent default on their loans, the 4,000,000 REA customers have not been recipients of charity, but bona fide borrowers of money. By providing area coverage,

REA could serve all farms. Use of the Rochdale concept of cooperatives made the recipients partners in a business venture and more likely to support REA.

It was the numerous benefits made possible by electrification, however, that accounted for the popularity and success of the agency. But the impact of electrification was not immediately visible because the benefits were available only as fast as families could equip their homes and farms with appliances and labor-saving machines. In three USDA surveys, practically all homes regardless of income level first had a radio and flat iron. High income farms had a greater diversity of equipment and a higher use of energy. The bulk of energy consumed on all farms for all purposes, however, was for household operations, amounting to 90 percent in some cases. Despite the difficulty of measuring the impact, changes in the rural lifestyle were quickly apparent.[11]

Incandescent lighting ended the use of lanterns and brightened the interior of the home, creating a warmer and more congenial atmosphere. Shadowy corners and dimly-lit hallways became a thing of the past. Leisure time was increased: TVA reported that electric lights added two to four waking hours per day; another study showed that ninety-one "8-hour days" per year were added to the waking hours of farm families. New lighting also encouraged reading and stimulated children's interest in schoolwork. The American Society of Electrical Engineers stated that incandescent lights reduced by 50 percent the amount of time required to perform chores with a lantern. More than anything else, however, electric lights meant deliverance from the dark. "Mother," a small boy said when his family acquired service, "I didn't realize how dark our house was until we got electric lights."[12]

The most popular convenience was indoor plumbing. A farm journal poll of subscribers reported that 7,233 respondents regarded running water as the best improvement in removing the drudgery and toil of farm life. More than one-half hour per day was saved with this new comfort. Indoor faucets eliminated the assortment of small tubs and pails that previously had cluttered the kitchen and back porch. An electric clothes washer saved twenty days per year of scrubbing laundry in a tub out of doors. Electric irons did not require a hot stove on which to be heated; they weighed less; they maintained a constant and even temperature; and they could be operated for only a few pennies per month. Popularity of

the iron was evident since it was one of the first items acquired. The outdoor privy disappeared as soon as indoor bathrooms were installed.
"Running water," wrote one observer, "provides health protection and
relief from drudgery. It is probably the most important single benefit
made possible by electricity."[13]

Refrigerators also promoted family comfort and health while improving
the appearance of the premises. At least one outdoor structure, either
the blockhouse or the smokehouse, was no longer essential to the farm
once refrigeration was available. Ironically modern cold storage was responsible for the decline in the number of home gardens, because fruits
and vegetables could be stored for several days inside the refrigerator,
and many housewives preferred to replenish the family supply of vegetables at the grocery instead of tending a garden. Southerners acquired
refrigerators at a higher rate than consumers in any other region. Diets
improved since perishable foods high in vitamins and protein could be
stored longer. Visitors at the pre-electrified farm table had noticed the
large amount of starchy foods that required little coolant, but visitors to
the modernized farm likely found little difference in the table offering
of their host and that in city homes. Refrigeration also reduced the high
incidence of food spoilage and staphylococcus food poisoning on farms.[14]

Particularly beneficial were the effects of electrification on health.
In a study of lighting and health, the United States Public Health Service
(PHS) reported that incandescent illumination protected eyesight, pro-
moted cleanliness, prevented accidents and had psychological benefits.
Approximately 5,000,000 persons per year sustained injuries from home
accidents, of which 150,000 were disabling and another 39,000 fatal. In-
sufficient light was listed as the cause of nearly half of these accidents.
While these statistics included urban residences in the United States, they
demonstrated, nonetheless, the importance of proper lighting in any home.
Squalid, half-lit houses have, the report added, a depressing impact on the
inhabitants, who "are bound to be affected . . . by its mentally unstimulat-
ing qualities."[15] No statistical evidence was available to demonstrate the
health value of indoor plumbing, but since "it was a well known fact,"
as the PHS reported, "that a great deal of the sickness occurring on the
farm . . . has been caused by polluted water . . . ," the installation of a
modern waste removal system and running water obviously had salutary
effects.[16] In 1935 Cooke had employed Mollie Ray Carroll, a PWA social

worker, to study the effect of indoor plumbing, and she reported that family hygiene improved remarkably, that housewives washed food, clothing, and linen more thoroughly.[17]

The secondary effects of a comfortable home environment were also known, for good health was, as the PHS reported, a "state of being in which all physical and mental processes function at their highest efficiency."[18] Peace and satisfaction were vital elements of health. While these comfort-giving influences of electrification were intangible, they were, nonetheless, real. Thus, improvement in the home went beyond the easily measured mortality and sickness rates.

Not all benefits of electricity were intangible; the radio ranked among the most popular new devices in the home. It brought information on weather, marketing, livestock sales, meetings and related items. Forewarning of a storm helped to avoid loss of crops. Each family had its favorite radio serials, news and entertainment programs. With the aid of radio, rural inhabitants became more aware of the ebb and flow of life in the world, and it enabled them to play a less passive role in an industrial society. Cultural-social differences between city and country were thus reduced, but in heightening the awareness of the cultural differences, radio encouraged youth to acquire a more sophisticated mode of life, and they flocked to the cities in search of satisfaction. In this sense the new instrument intensified migration to the city, the antithesis of what the early proponents of electrification had hoped would happen. In its overall effect, however, radio contributed to a sense of conformity and outlook.[19]

Just as the use of electricity added to the comfort of the home and contributed in varying degrees to relief from drudgery, it had the same effect on farming operations. Lights were nearly always the first outdoor use of electricity, but the fractional horsepower portable motor was the most versatile electrical aid. Lightweight and easily carried from place to place, it performed jobs requiring small amounts of power, such as operating a grinding wheel, drill press, wood saw, fan or other small equipment. With these devices the expense of farm labor was cheaper than when performed by hand.[20]

The effects of electrification were most conspicuous in the South. In the Piedmont, USDA reported the conversion of cotton farms into dairies once electricity was available. Poultry farming also spread; by 1955 the South produced 69 percent of the broilers consumed in the United States compared with 27 percent a generation earlier. One-eighth of the South's total income in 1955 came from the sale of poultry and eggs. As an astonished Georgia farmer expressed it: "Who'd ever thought a dad-blamed chicken would scratch cotton off the land?"[21] In 1956 *Progressive Farmer* reported an increase of 344 percent in southern rural living standards since 1940 compared with 234 percent for the whole nation. For that improvement the magazine gave credit to the "power revolution," meaning the greater comfort and convenience of home life due to electrification.[22]

It was for these reasons that the South had such an important influence on public electrification. Nowhere was it needed more and nowhere were its benefits greater. From the beginning of research in the 1920s through the half-century mark in 1950, southerners were fiercely devoted to rural electrification. It was no accident that the only significant CREA experiment occurred in Alabama, and it was no wonder that three of the best-known leaders, Rayburn, Ellis and Rankin, were from southern states. The abundance of streams for power plants also explained the southern push for public development of energy sources for rural use. Living in preindustrial conditions, but surrounded with the resources to modernize their life, southerners saw federal intervention as the answer. They knew, too, that since their region had such a high proportion of small, poor farms, power companies would not enter the rural market and public action was their only hope.

The electrification of agriculture was, however, a collective effort in all regions. Senator George Norris of Nebraska was the "father of the TVA," the greatest power development in the South. From Pennsylvania came the recognized founder of REA, Morris Cooke. President Roosevelt was a product of the Hudson River Valley, and his successor, Harry Truman, another enthusiastic supporter of REA, hailed from the Midwest. This national character was evident in the role of the Farm Bureau and Grange, two widely based agricultural organizations. Prior to REA's start, farmers in different regions had organized cooperatives, demonstrating again the broad geographic scale of the need for electric service and the determination to have it.

Predictions about the use of electricity generally came true. Rural inhabitants quickly adapted it to their needs. Area coverage, as Cooke had prophesied, proved feasible with the use of graduated rates and the availability of small appliance loans to consumers. Predictions about the use of electricity ending farm to city migration, however, proved wrong. Electrical technology displaced workers and encouraged migration even while it made life more attractive and kept people on the farm.

For those who chose to remain at home, electricity removed the drudgery of labor and brought the comforts of the city to them. With electricity the farmer replaced animal power with machines, and housewives enjoyed modern conveniences. Through the radio rural folk had broader cultural horizons and no longer felt inferior to their city kin. Even personal satisfaction improved; one farm wife spoke for all when she wrote: "I am enjoying life more because I have more time to spend visiting my friends, studying and reading, and doing the things that make life richer and fuller."[23]

11
IDEOLOGY OF
RURAL
ELECTRIFICATION

Extending electricity to American farms was achieved after a long
struggle by agricultural and political interests that exhibited ideological
inconsistency in their endeavor. Most supporters had come from non-
electrified farms and remembered the drudgery and pathos of preindustrial
life. But several of the most important figures, Roosevelt and Cooke, for
example, were products of a comfortable environment. Only one common
denominator was apparent among them: they espoused the Jeffersonian
ideal of agrarian values, seeing electrification as a means of restoring rural
life and replenishing, so to speak, the nursery stock of American leaders
and source of democratic values.[1] Their lack of consensus was due to
their geographic and cultural diversity and their representation of a con-
stituency, the rural population of the United States, whose views varied
on electrification as well as on more general political and economic questions. By 1935 when REA was started, the leaders agreed service could
be extended to farms only by government action, but they disagreed
sharply over the extent of such action.

   There was inconsistency at the end of World War I in the initial efforts
to serve farms. Electric service was just then being recognized as a necessary ingredient for modern living, and the political climate was shifting
from the activist influence of the progressives to the conservative, laissez-
faire philosophy associated with business. It was the power industry that
showed the first interest in rural electrification when it created CREA.

That experimental undertaking proved, of course, to be a failure, but it had the blessings, and indeed the participation, of the American Farm Bureau Federation. It would be wrong to say the latter should not be included among the precursors of public electrification since in later years it fully endorsed REA and had a part in the decision to use cooperatives. Furthermore, CREA had the endorsement of prominent rural magazines and newspapers which regarded private enterprise as the appropriate instigator of service.

Diametrically opposed was the approach taken by progressives such as George Norris and Judson King. To their way of thinking, the extension of rural service would never occur except by public means; it was foolish to expect the "power trust" to act responsibly and develop the rural market. In classical progressive fashion, this camp sought to expand democracy and promote social betterment through massive public power systems. They championed public electrification in Canada and Europe, referring to it as an example of government regulation and even direct competition that was necessary to break the grip of corporate wealth on the American consumer.

For Norris, the electrical industry should be placed on trial and found guilty for not making a full-scale commitment to connect farms and rural homes. The handful of successful electric cooperatives operating in the 1920s reinforced Norris's contention that public electrification was feasible. Use of the cooperative fitted into the progressive mold of thinking, for progressives regarded it as a solution for other agricultural problems. In regard to rural electrification they were willing to go beyond regulation and compete with corporations, an idea in agreement with their general concept about public power.

It might appear, however, that although the power industry vehemently opposed Norris, it was in rather close agreement with the progressive ideology of Morris Cooke, "father of the REA." A superficial understanding of the evidence would encourage such an interpretation, one in keeping with similar interpretations of business participation in progressive reforms.[2] Giant Power encouraged private development of power plant sites next to coal mines and offered the power of eminent domain for acquiring transmission line rights of way. Utility companies were encouraged to serve rural homes by taking advantage of the states' services and jurisdictional authority whenever necessary. Only when the respective company refused service could a cooperative be formed. Such a proposal differed sharply with Norris's preference for a wholly public operation. Conceivably the Giant Power plan could have extended the electrical industry's "grip" on rural inhabitants, enabling it to expand its horizons with governmental approval, and giving the industry greater security and a defense from public power enthusiasts such as Norris.

Similarities between Giant Power and SuperPower would further encourage suspicion of an ideological alliance of progressive and business thought. The latter plan also called for mine-mouth generating plants, pooling of energy supplies and huge transmission lines carrying current across state boundaries. Hoover was the chief spokesman for SuperPower and the resemblance of his background and views to Cooke's was striking. Both were engineers who believed in social responsibility and efficiency in government. They saw objectivity and selflessness as keys to economic progress and accepted redistribution of wealth as a necessary step for social improvement. Hoover described SuperPower as a chance to spread technological benefits to all citizens and lift the burden of hardship and drudgery from their shoulders. In 1920 Cooke had urged his friend to run for president, and when Hoover set up the Northeastern SuperPower Committee in 1921, he asked Cooke to serve on it.

It was the question of rural electrification that revealed the difference in these two plans and which drove Hoover and Cooke into opposite camps. Whereas service for the farmer was the heart of Giant Power, SuperPower made no reference to it. Hoover told Cooke that he saw the "agricultural problem as one of first getting our primary system onto right lines," meaning that utilities would voluntarily connect farms after developing the industrial market. SuperPower demonstrated Hoover's belief in trade associations and "pools," fitting perfectly his concept of corporate voluntarism and ultimately his "American individualism." He further opposed the recommendations in Giant Power to encourage rural service.

Cooke and Pinchot became suspicious of SuperPower and interpreted it as a limited plan aimed at promoting corporate profits under the pretense of scientific and engineering objectivity. Giant Power on the other hand sought to provide technological rewards to the public. "Not all progressives," according to one writer, "made this distinction in terms," which could in this case encourage an interpretation of Hoover as a progressive.[3] But he stood in sharp contrast to their proposals for electrification, putting great hope on the voluntary willingness of utility companies to serve the rural market. By such action, Hoover stifled his own hopes of material independence for Americans, for without electricity and its benefits, how could rural families be independent? In regard to electrification, Hoover steered the path of voluntarism traditionally associated with him. He refused to use the Giant Power plan to his advantage, even if it conceivably could have strengthened corporate might since Cooke preferred private development.

When he became president, Hoover continued to resist plans for electrifying agriculture unless they fitted his notion of corporate voluntarism. In 1931 Cooke, sensing a chance to win presidential approval of a program under the guise of stimulating employment, convinced Colonel Arthur Woods, chairman of the president's Commission on Employment, to establish an "emergency, non-profit organization . . . in full cooperation with the electrical industry."[4] Here was Hoover's chance either to promote a combination of government and business by vastly expanding the power industry's area of operations, or to prove himself a progressive by using Cooke's idea in the name of socioeconomic improvement. He flatly rejected Cooke and held to the belief that electrification should be achieved only by individual power companies acting independently and resorting to sound business principles. The former president's position on the subject would not alone be enough to judge him a conservative, progressive or economic syndicalist, but his action, which conformed to the interpretation of a conservative Hoover, must be taken into consideration.[5]

It would have been wrong to make ideological enemies of Norris and Cooke. Each emphatically insisted that strong government intervention was necessary to electrify rural homes and farms; one had no faith in the utilities while the other wished to make them subject to stiff and inescapable regulations. They worked closely on other matters, regularly supplying one another with information and other forms of assistance. The utilities' resistance to Cooke's Giant Power should remove any suggestion that progressives and business were ideologically alike on the subject of rural electrification.

In the meantime, Norris kept pounding away, insisting that power companies had no intention of serving farmers either independently or through joint governmental programs. The 1930 Census confirmed his claim, showing no appreciable gain in the number of electrified farms except in the Pacific states where special conditions had, of course, attracted the

utilities. Patience with the power industry had ended with the poor showing by CREA and the lack of any action to come out of the SuperPower
concept after Giant Power had been relegated to a dusty shelf. These
failures by private enterprise brought a greater acceptance of public
electrification, an attitude manifest by the creation of Public Utility
Districts in Washington and the steps taken in the Carolinas to inaugurate
wholly public systems. Roosevelt's endorsement of public power in 1932
convinced Cooke to break his ties with the Republicans, and he endorsed
the Democratic candidate for president. Public electrification had esca-
lated because there was no visible alternative. Indicative of this development was the Farm Bureau's switch calling for federally assisted cooperatives.

When Cooke launched his campaign within the administration for a
nationwide program, however, he demonstrated much inconsistency,
indicating that although public electrification had become more widely
accepted, reservations were still extant. He had to recognize and accommodate the indeciveness among officials over the role of the power
industry. He suggested joint programs and wholly public ones. Through
TVA a public program was undertaken on a limited scale, but until the
results were clear he had to consider industry as a partner. But Cooke
still leaned toward business participation by his own accord for when
Roosevelt created REA in 1935, he immediately went to the utilities.[6]

Cooke urged the utilities to join REA as partners, offering massive
capital funds if they promised area coverage and rate reductions. He even
delayed REA by several months because of his hopes for industry participation. When he approved the first loans to cooperatives, he did so reluctantly because he feared that farmers would prove unable to handle
the details of operation. By this time, Cooke was on the political right
since congressional members such as Norris and Rankin, along with the
Farm Bureau and National Grange, pressured for the use of cooperatives.
Only when the power industry refused Cooke's offer did he abandon his
hope for a joint program.

Cooke's attempt to establish a role for business using tax dollars follows
the interpretation of the New Deal as a major event in the development of
a syndicalist nation. Rural inhabitants might also suffer "double economic
jeopardy" by financing their own electric lines through REA while having
to pay the profits of the respective utility companies serving them.[7] In
regard to REA, interpretation of the New Deal as promoter of syndicaliza-

tion conflicts with the facts in two respects. The Public Utility Act of 1935 passed in the midst of Cooke's negotiations with the power industry was a real attempt to reduce the syndicalization that had already occurred. Only resistance by the industry kept the measure from going further. And Cooke's demands for low rates and area coverage would have kept the participating companies responsible to the public. It was his concern for the poorest farms that drove them away. For these reasons it would not be accurate to say REA recklessly pushed the government into the power business or reinforced the utilities' grip.

Underlying Cooke's effort for a joint program was his old belief in efficiency going back to his association with Frederick Taylor. The former had a reputation in the New Deal for always seeking the least costly method of achieving a particular goal. Resource use in his opinion should be minimized whether for the purpose of social welfare or corporate profit. It was this consideration, efficiency, that attracted him to the electric companies that were already equipped and experienced.

The triumph of the cooperative appeared to answer the question of a joint program with business, but such was not the case. In 1936, when Norris moved to establish the REA by congressional authorization, preference for utility participation reappeared. Congressional opposition to excluding the utilities was so strong that the REA bill almost failed. Rayburn was the principal proponent of industry's participation; he and Norris compromised, knowing, however, that the electric companies would not apply for REA funds. Rayburn's success in gearing the REA rate of interest to fluctuating economic conditions was more significant because that provision kept REA from operating in an artificially created market and from establishing itself as a self-contained bureaucracy unanswerable to anyone except its own supporters. Despite any inferences, Rayburn was not a handmaiden of the power industry; his belief in public electrification was evident in his stance on the Slattery incident.

The Slattery incident proved to be an important battle over the definition and purpose of REA. Ellis and the contingent of NRECA supporters wanted to use the agency as an instrument to help the rural poor. They hoped to lower the expense of co-op operations by reducing insurance costs and other business items. More important, NRECA sought to organize the cooperatives into a collective body to defend themselves against the power industry and also pressure Congress for fatter REA appropriations.

Old-line progressives such as Slattery and King opposed NRECA. Norris at first supported Slattery because he saw the new lobby as a ploy by would-be politicians and ambitious bureaucrats to seize control of REA. Even Cooke once mistakingly thought it was allied with the electrical industry. Confusion was apparent in the allegations made in regard to NRECA's insurance proposals, and some differences existed over REA's labor policy. At the heart of the dispute, however, was the extension of REA into power generation for much of the thrust of the attack against Slattery came from rural electric associations in the states wanting REA to build power plants. Rayburn supported both subsidization and power generation when he realized they were required to extend service to all classes of farmers. Norris lost his bid for re-election in the midst of the fracas, but when he realized the true intentions of NRECA, he lent his name to it. Wickard's conflict with Slattery had, according to Bledsoe, "no conservative, liberal or ideological significance."[8] Wickard saw Slattery as a very sick man who, despite his reputation as an effective liberal, was an inept administrator.

Since most REA supporters ultimately agreed with Ellis, the incident could be interpreted as nothing more than a tragic failure for Slattery whose obstinance shoved him into oblivion. But the episode had significance: it represented the emergence of a new generation anxious to move faster and also into new areas. The impatience and to some extent the belligerence of youth collided with the slower pace of age. Some old-line progressives had wanted to work jointly with business, and some had envisioned REA as a distributor of electricity rather than a producer. NRECA wanted the agency and its supportive apparatus to turn its back forever on business and take on the qualities of an organized special interest, the antipathy of the older generation's dreams of an independent and small agency whose interests extended only to the individualistic yeoman.

Successful passage of legislation in 1944 for subsidized REA loans plus the growth of power generation indicated the emergence of liberal ideologists ready and willing to expand public electrification. When Ellis and Rankin called for the "little TVA's" they expressed that view. Beginning with the construction of dams in the Southwest, Ellis saw a chance to re-build farms on a large-scale, extending not only electricity but also promoting conservation, reforestation, fertilizer production and other benefits. Massive federal aid would, he hoped, lift the rural classes out of poverty.

This call for "little TVA's" constituted the most radical ideological stance taken after World War II and was a descendent of Norris's earlier call to end business dominance in the power industry.

Rayburn led the more conservative faction that believed the worst of the socioeconomic conditions had passed during the war. Poverty still existed, they admitted, but not on a scale warranting another wave of large-scale intervention. Rayburn concentrated on the energy shortage, insisting it was unfinished business, and he knew, furthermore, that the conservative postwar climate left no chance for "little TVA's."

Since the postwar fight for energy supplies in the South was so bitter, Rayburn's stance staved off a disaster. "The utilities were pushing their views with a vigor unknown since the days when utility propaganda was being planted in school textbooks," according to one reporter.[9] Only Rayburn's clout with Truman and his prestige in the House enabled him to save the dams from private control. Ellis's more radical position did not stand a chance in view of the tough opposition.

Rayburn's decision to make partners of utilities in the Southwest included more than a realization of the political facts of the time. He wanted to blend public and private facilities for the sake of maximum efficiency. Ellis's opposition to these arrangements again demonstrated an ideological difference, but he could not match Rayburn's strength. Duplication of private lines and generation plants was a wasteful use of resources in Rayburn's view; better to cooperate by exchanging power, he would say, in order to guarantee a supply of energy to the cooperatives for a longer period of time. Shortage of water in the Southwest was partly a reason for exchanging power, but Rayburn blocked all moves for SPA to build steam-generating plants. In this sense his thoughts resembled Cooke's who believed cooperation was more efficient. Finally, Rayburn questioned the wisdom of erecting "superstates" or gigantic agencies independent of the usual controls as envisioned in some of the river basin proposals.

Public electrification failed to satisfy those hoping to see some restraint in the growth of corporate wealth. The reduction of rural rates of private suppliers, although beneficial to the public, in no way affected the strength of the electrical giants. More radical constraints were prevented by political exigencies, namely, the power industry opposition. Ideological agreement among the leaders would not have resulted in

greater restraints on the electrical industry because political conditions
were simply too unfavorable. The major handicap of public electrification was not, therefore, ideological disagreement, but stiff and effective
resistance by the private interests.

REA still encouraged economic decentralization since it formed more
than 1,000 cooperatives serving 5,000,000 families with electricity, a
modern example of grassroots involvement. It encouraged individualism
by equipping rural inhabitants with the means to free themselves of the
economic and social restraints of a preindustrial lifestyle. Public electrification promoted individualism while not checking business growth since
it served a market the private interests chose to ignore.

In the final analysis, electrification did not restore the rural way of
life; it failed to check the migration into cities or stop the decline of
small family farms. Use of electricity, however, greatly improved home
life and quickened the modernization of farming operations. The ultimate effect was not a revival of an older order but entry into the new.
As the United States became an industrialized society with a high percentage of urban inhabitants, electrification enabled the rural sector
to keep pace in a social and cultural sense and also retain its place in the
world as supplier of food and fiber.

NOTES

Introduction

1. A. M. Daniels, "Electric Light and Power in the Farm House," *Yearbook of the United States Department of Agriculture 1919* (Washington, D.C., 1920), p. 223.

2. Louisan Mamer, "Electricity Pays Its Own Way in the Rural Home," typescript (March 11, 1952), National Agricultural Library, p. 10; REA, Interbureau Coordinating Committee on Rural Electrification, "Present Uses of Electricity in Rural Areas," typescript (1941), p. 1; Mercer Green Johnston Papers, Library of Congress, Manuscript Division.

3. Victor M. Ehlers and Ernest W. Steel, *Municipal and Rural Sanitation* (New York, 1943), p. 372; J. H. Mason Knox, Jr., "Reduction of Maternal and Infant Mortality Rates in Rural Areas," *American Journal of Public Health and the Nation's Health*, 25 (January 1935): 72; R. G. Upton, "Incidence and Severity of Hookworm Infestation in East Texas," ibid., 26 (September 1936): 924-26; James H. Cassedy, "The Germ of Laziness in the South, 1900-1915: Charles Wordell Stiles and the Progressive Paradox," *Bulletin of the History of Medicine*, 45 (March-April 1971): 159-69.

4. Paul de Kruif, "The Rise and Fall of Pellagra," *Readers Digest*, 31 (September 1937): 75-78; Dickson Hartwell, "The Miracle of Dr. Spies," *Colliers*, 121 (January 31, 1948): 26; Wilson G. Smillie, *Preventive Medicine and Public Health* (New York, 1956), p. 405.

5. H. G. Birch, "Malnutrition, Learning and Intelligence," *American Journal of Public Health*, 62 (June 1972): 773-84.

6. Bureau of the Census, *Fourteenth Census of the United States, Agriculture*, 5 (1922): 23, 512-14; ibid., *Fifteenth Census of the United States, Agriculture*, 4 (1932): 10.

7. University of North Carolina, *News Letter* (October 29, 1919), p. 1.

8. "Speech of Gifford Pinchot Before National Electric Light Association," typescript, (May 21, 1924), Morris L. Cooke Papers, Box 188, Franklin D. Roosevelt Library.

1: The Farmer and the Electrical Industry

1. National Electric Light Association (NELA), *Report of the Committee on Electricity in Rural Districts*, pamphlet (May 29-June 2, 1911), p. 3.

2. NELA *Bulletin*, 5 (1912): 301; Royden Stewart, "Rural Electrification in the United States; The Pioneer Period, 1906-1923," *Edison Electric Institute Bulletin*, 9 (September 1941): 382-83.

3. NELA *Proceedings* (1921), pp. 732, 851-52.

4. Ibid. (1923), p. 46.

5. Ibid. pp. 45-46; Stewart, "Rural Electrification," p. 386. According to O. M. Kile, long-time officer in the Farm Bureau, these initial meetings grew out of an attempt by officers of the Farm Bureau and NELA to get a reduction of rural rates in Illinois. Unsuccessful, they took the problem to their respective organizations, thus, providing the impetus for the founding of CREA. O. M. Kile, *The Farm Bureau Through Three Decades* (Baltimore, 1948), pp. 78-80.

6. Harry Slattery, *Rural America Lights Up* (Washington, 1940), p. 16. For the original statement see NELA *Proceedings* (1923), p. 67.

7. NELA *Bulletin*, 11 (March 1924): 146-47; W. M. Jardine, "Some Difficulties and Possible Remedies Standing Before Rural Electrification," NELA *Bulletin*, 41 (July 1925): 446-47; Marquis Childs, *The Farmer Takes a Hand* (New York, 1952), p. 39.

8. Slattery, *Rural America Lights Up*, p. 5; Childs, *Farmer Takes a Hand*, pp. 42-43; Royden Stewart, "Bringing Power to the Farm: Early Development Years," *Public Utilities Fortnightly*, 27 (May 8, 1941): 579-87; F. G. Gausemeier to Franklin D. Roosevelt, March 17, 1936, Franklin D. Roosevelt Papers, Box 1, OF 1570, Franklin D. Roosevelt Library; L. K. Richardson, *Wisconsin REA: The Struggle to Extend Electricity to Rural Wisconsin, 1935-1955* (Madison, 1961), pp. 6-7.

9. NELA *Proceedings* (1924), pp. 64-65; E. A. White, "Relation of Electricity to Agriculture Committee," ibid., pp. 80-81; Royden Stewart, "Bringing Power to the Farm: National Development 1924-1935; The CREA," *Public Utilities Fortnightly*, 27 (May 22, 1941): 651-63; Childs, *Farmer Takes a Hand*, pp. 39-42; Twentieth Century Fund, *Electric Power and Government Policy* (New York, 1948), pp. 440-41.

10. NELA Rural Electric Service Committee, *Progress in Rural and Farm Electrification*, 1921-1931, NELA Publication no. 237 (August 1932); CREA, *Tenth Annual Report 1933*, typescript, pp. 1-20, Alabama Power Company Library.

11. NELA *Bulletin*, 11 (January 1924): 23-45; Slattery, *Rural America Lights Up*, pp. 18-19; Stewart, "Bringing Power to the Farm: National Development," p. 654.

12. *Ibid.*, pp. 654-655; NELA *Bulletin,* 11 (May 1924): 216; ibid., (November, 1924), 667-670; ibid., 18 (July 1931): 421; Childs, *Farmer Takes a Hand,* p. 42.

13. Slattery, *Rural America Lights Up,* p. 19.

14. Alabama Power Company, *Annual Report,* 1923, p. 9; Thomas W. Martin, *The Story of Electricity in Alabama* (Birmingham, 1953), pp. 74-78; quote in NELA *Bulletin,* 12 (September 1925): 554. In the words of one writer the Alabama CREA was "lost at sea" when they began. See Katherine Grimes, "Light Comes to Alabama," *Southern Agriculturalist* 59 (September 15, 1929), p. 5.

15. E. C. Easter, "History of Rural Electrification in Alabama," *Alabama Agricultural Engineer,* 1, Supplement (January 1954): 5; CREA *Bulletin* 1 (November 1924): 3; E. C. Easter to D. Clayton Brown, interview, June 17, 1969, Birmingham, Alabama.

16. Edison Electric Institute, *Bulletin,* 2 (January 1934): 9; M. N. Beeler, "Why Alabama Leads in Electrifying Farms," *Capper's Farmer* 38 (April 1927), p. 8; "Progress Report of Project in Rural Electricity," Department of Agricultural Engineering, Alabama Polytechnic Institute, 1925, typescript in Alabama Power Company Library; quote in NELA *Bulletin,* 12 (September 1925): 554.

17. Ibid.; CREA *Bulletin,* 1 (November 1924): 4.

18. Agnes E. Harris, "What Electricity Meant to the Farm," *Powergrams, Alabama Power Company,* 6 (October 1926): 17; Beeler, "Why Alabama Leads," pp. 7-8; NELA *Bulletin,* 12 (April 1924): 235; ibid., (October 1924): 605; quote in ibid., 13 (September 1926): 773.

19. M. L. Nichols, "Research Work and Rural Electrification," *Powergrams, Alabama Power Company,* 6 (October 1926): 11; CREA *Bulletin,* 2 (April 1926): 6.

20. Quote in ibid., p. 12; E. C. Easter to D. Clayton Brown, Interview, June 17, 1969.

21. M. J. Funchess, "The Agricultural College and Rural Electrification," *Powergrams, Alabama Power Company,* 6 (October 1926): 5.

22. South Carolina Power Rate Investigating Committee, *Report on the Electric Utility Situation in South Carolina* (Columbia, 1931), pp. 457-59; Slattery, *Rural America Lights Up,* p. 19. Statistics varied on Alabama farms receiving central station service. The U. S. Census reported 2.7 percent, but CREA reported 4.8 percent. In either case the percentage was small. See Bureau of the Census, *Fifteenth Census of the United States, Agriculture,* 4 (1932): 518; NELA *Bulletin,* 19 (September 1932): 525.

23. CREA *Bulletin,* 2 (May 1926); NELA *Bulletin,* 12 (September 1925): 567; Allen J. Saville, "The Steady Demand for Rural Electrification: The Virginia Plan," *Public Utilities Fortnightly,* 16 (September 12, 1935): 312-13; REA Memorandum, Mercer Johnston to Morris Cooke, July 24, 1936, Morris L. Cooke Files, Box 7, National Archives, Record Group 221; NELA *Bulletin,* 14 (June 1927): 370-71.

24. *Fifteenth Census of the United States, Agriculture,* pp. 12, 508, 518; NELA *Bulletin,* (September 1932): 524-28; CREA *Bulletin,* 6 (June 1931); Slattery, *Rural America Lights Up,* p. 21; quote in Memorandum, Johnston to Cooke.

25. NELA *Proceedings* (1923), p. 46.

26. Quote in Childs, *Farmer Takes a Hand*, p. 42; Richardson, *Wisconsin REA*, p. 7; Richard Lowitt, *George W. Norris: The Persistence of a Progressive, 1913-1933* (Urbana, Ill., 1971), p. 333.

27. Stewart, "Bringing Power to the Farm: Early Development Years," p. 586.

28. John Bauer and Nathaniel Gold, *The Electrical Power Industry: Development, Organization and Public Policies* (New York, 1939), p. 120.

2: The Alternatives: Cooperatives and Public Power

1. "List of Pre-REA Electricity Distribution Cooperatives," Judson King Papers, Box 79, Library of Congress, Manuscript Division; Harcourt A. Morgan, "Rural Electrification: A Promise to American Life," *Transactions Third World Power Conference*, 8 (Washington, 1938): 796-98; David C. Coyle, *Electric Power on the Farm*, REA Pamphlet (Washington, 1936), pp. 60-62; Clyde T. Ellis, *A Giant Step* (New York, 1966), pp. 33-34; Jerry Voorhis, *American Cooperatives* (New York, 1961), p. 56.

2. REA Report, Udo Rall to Jacob Baker, May 16, 1935, Morris L. Cooke Files, Box 6, National Archives, Record Group 221; Harry Slattery, "Rural Electrification for the Millions," typescript, King Papers, Box 79.

3. REA Report, Rall to Baker.

4. Ibid.; Albert B. Gerber, "Federal Administration of Rural Electrification," Masters Thesis, George Washington University, 1942, p. 14.

5. REA Report, Rall to Baker; NELA *Bulletin*, 2 (1915): 805; quote in REA Memorandum, John Carmody to Clay Cochran, January 26, 1955, John M. Carmody Papers, Box 83, Franklin D. Roosevelt Library.

6. Morgan, "Rural Electrification," pp. 785-99.

7. Marquis Childs, *Sweden, The Middle Way* (New Haven, 1936), p. 79.

8. Ibid., pp. 75-83.

9. REA Reports 1934-1935, King Papers, Box 78; Harold Evans, "The World's Experience with Rural Electrification," *Annals*, 118 (March 1925): 30-42. For the power industry point of view on foreign rural electrification, see NELA *Proceedings* (1925), pp. 72-76.

10. Childs, *Sweden, The Middle Way*, pp. 75-89; E. A. Stewart, "Rural Electrification in Europe," Journal Series Paper no. 661, Department of Agriculture, University of Minnesota (1927), Alabama Power Company Library; NELA *Proceedings* (1925); Gerber, "Federal Administration of Rural Electrification," pp. 14-16.

11. Evans, "World's Experience with Rural Electrification," pp. 35-36; *Rural Electrification News*, 1 (April 1936): 7.

12. Richard Lowitt, *George W. Norris, The Persistence of a Progressive, 1913-1933* (Urbana, 1971), p. 263.

13. Ibid., pp. 197-209; "Interview with Senator Norris RE: Muscle Shoals History," typescript, (August 31, 1935), Box 52, King Papers.

14. *Congressional Record*, 69th Congress, 1st Sess., Vol. 67, pt. 1, pp. 332-61. For a discussion of the Wyer pamphlet, see Lowitt, *George W. Norris*, 264.

15. Lowitt, *George W. Norris*, p. 264.

16. For a discussion of PUDs, see Michael Knight Green, "A History of the Public Rural Electrification Movement in Washington to 1942." Doctoral dissertation, University of Idaho, 1967.

17. Judson King, *The Conservation Fight: From Theodore Roosevelt to the Tennessee Valley Authority* (Washington, 1959), p. vi.

18. Ibid., p. xv.

19. Ibid., p. v.

20. *New York Times*, July 5, 1958, p. 17.

21. E. A. Stewart, "Electricity in Rural Districts Served by the Hydro-Electric Power Commission of the Province of Ontario, Canada" (1926), p. 6, Alabama Power Company Library; Judson King, *National Popular Government League (NPGL), Bulletin* no. 127, (April 10, 1929), pp. 23-26; quote in King, *The Conservation Fight*, p. 191.

22. Quote in George Norris, *Fighting Liberal* (New York, 1945), p. 318. For expressions of concern over farm to city migration, see Howard Odum, *American Social Problems* (New York, 1939), pp. 84-104; E. R. Eastman, *These Changing Times* (New York, 1927), pp. 220-28; Harry E. Moore and Bernice M. Moore, "Problems of Reintegration of Agrarian Life," *Social Forces*, 15 (March 1937): 384-90.

3: Morris L. Cooke and Giant Power

1. Kenneth Trombley, *The Life and Times of a Happy Liberal: Morris Llewelyn Cooke* (New York, 1954), pp. 1-30.

2. Ibid., p. 45; Jean Christie, "Morris Llewelyn Cooke: Progressive Engineer." Doctoral dissertation, Columbia University, 1963, pp. 31-32.

3. Quoted in Jean Christie, "Giant Power: A Progressive Proposal of the Nineteen-Twenties," *Pennsylvania Magazine of History and Biography*, 96 (October 1972): 484.

4. Copy of speech in Morris L. Cooke Papers, Box 188, Franklin D. Roosevelt Library.

5. Morris L. Cooke to W. S. Wise, October 23, 1924, Cooke Papers, Box 195. Material relating to the Giant Power Survey is found in the Gifford Pinchot Papers, Philip P. Wells Correspondence, 1923-1927, Boxes 1696-1699, Library of Congress, Manuscript Division.

6. John H. Gray, "Giant Power: The Giant Power Report of Pennsylvania," *National Municipal Review*, 15 (March 1926), 165-72.

7. George Morse, "Rural Electrification," *Giant Power Survey Report* (Harrisburg, 1925), pp. 117-40.

8. Morris Cooke to Philip Wells, July 1, 1924, Cooke Papers, Box 189.

9. *Annals*, 118 (March 1925): viii.

10. John Gray, "Giant Power," pp. 168-169.

11. Gray, "Giant Power," p. 169; Judson King, *National Popular Government League Bulletin no. 127*, (April 10, 1929), p. 18.

12. Christie, "Morris Llewelyn Cooke," pp. 94-96.

13. Gray, "Giant Power," p. 170.

14. King, *National Popular Government League Bulletin no. 127*, p. 18; quote in Thomas McCraw, *TVA and the Power Fight, 1933-1939* (Philadelphia, 1971), p. 2.

15. "A League for Super-Power," *The Literary Digest*, 67 (November 13, 1920): 30-31; Morris L. Cooke, "The Long Look Ahead," *The Survey*, 51 (March 1, 1924): 600-04; Herbert Hoover, Address to National Electric Light Association, May 19, 1922, typescript, Herbert Hoover Papers, COF 290, Herbert Hoover Library.

16. James H. Collins, "Sober Sense About Super-Power," *The Nation's Business* 12 (March 1924): 21-23; Cooke, "The Long Look Ahead," pp. 601-604; quote in *Annals*, pp. viii-ix.

17. Trombley, *The Life and Times of a Happy Liberal*, pp. 94-96; quote in Hoover to Super Power Conference, October 13, 1923, typescript, Hoover Papers, COF 290.

18. Hoover to Morris Cooke, December 13, 1923, Hoover Papers, COF 98; quote in *Power*, 64 (November 30, 1926): 845.

19. Cooke, "The Long Look Ahead," p. 602.

20. Hoover to Cooke, April 2, 1924, Hoover Papers, COF 290.

21. *The Survey*, 51 (March 1, 1924); Samuel Gompers, "Giant Power—Its Possibilities, Potentialities, and its Administration," *American Federationist*, 30 (December 1923): 570-71; ibid., (August 1924): 621-28; *New York Times*, July 2, 1925, and February 9, 1926; *Sacremento Daily Union*, July 6, 1924; NELA *Bulletin*, 12 (March 1925): 164-67; *Congressional Record*, 69th Congress, 1st Session, Vol. 67, pt. 1, pp. 341-42; "Senator Norris' Reference to Super Powers," typescript, (September 23, 1924), Hoover Papers, COF 290; Henry L. Stimson to Hoover, January 13, 1925, ibid.

22. *Giant Power: Proceedings before the Committee on Corporations of the Senate and the Manufacturers Committee of the House of Representatives, being a Joint Hearing on Senate Giant Power Bills, Numbers 32, 33, 34, 35, 36 and 37* (Extraordinary Session of 1926), pp. 24-32, (quotation p. 34), 47, 54.

23. Ibid., pp. 80-83.

24. King, *Bulletin* no. 127, pp. 20-21; *Giant Power: Proceedings*, p. 79.

25. Cooke to George Morse, Judson Dickerman, O. M. Rau, September 19, 1924, Cooke Papers, Box 189; Cooke Memorandum, September 24, 1924, Pinchot Papers, Box 1698; *Annals*, pp. 52-59.

26. *Philadelphia Patriot*, February 14, 1927, Pinchot Papers, Box 1698.

27. Stevenson W. Fletcher, *Pennsylvania Agriculture and Country Life, 1840-1940* (Harrisburg, Pa., 1955), pp. 62-65. Wells to Wayne Canfield, July 27, 1926. Quote in Wells to Pinchot, October 16, 1926; Wells to J. T. Worthley, October 19, 1926—all Pinchot Papers, in Box 1698; Cooke to W. S. Wise, May 8, 1930, Cooke Papers, Box 125.

28. Quoted in Frank Freidel, *Franklin D. Roosevelt, The Triumph* (Boston, 1956), p. 101.

29. Quoted in Marquis Childs, *The Farmer Takes a Hand* (New York, 1952), p. 51.

30. Cooke to Hoover, October 20, 1931, and Lawrence Richey to Cooke, October 21, 1931, Hoover Papers, Box 394; Arthur M. Schlesinger, *The Politics of Upheaval* (Boston, 1960), p. 381; Cooke to Arthur H. Woods, November 25, 1930, Cooke Papers, Box 230.

31. Cooke to Robert M. LaFollette, December 7, 1931, Cooke Papers, Box 121; quote in *Congressional Record*, 72nd Congress, 1st Session, Vol. 75, pt. 9, p. 10314. For Pennsylvania affairs, see Cooke correspondence with W. S. Wise, 1925-1933, Cooke Papers, Box 230; for South Carolina affairs, see Cooke Correspondence with R. A. Meares, 1930-1933, Cooke Papers, Boxes 121-122, 128-129 and 131. Cooke told George Norris that rural lines could be included alongside new highways, an idea he probably got from Meares in South Carolina. See Cooke to Norris, June 2, 1932, George W. Norris Papers, Tray 80, Box 7, Library of Congress, Manuscript Division.

32. Childs, *The Farmer Takes a Hand,* pp. 49-50; Trombley, *The Life and Times of a Happy Liberal,* p. 110; Cooke to John M. Walker, January 25, 1932, Cooke Papers, Box 124; W. K. McClenehen to Cooke, October 5, 1931, Cooke Papers, Box 131.

33. Morris L. Cooke, *What Electricity Costs in the House and on the Farm* (New York, 1933), pp. 75-117 (quotation p. 103); Cooke to Floyd L. Carlisle, October 15, 1932, Cooke Papers, Box 127; Cooke to Harold B. Johnson, February 3, 1933, Cooke Papers, Box 128.

34. Cooke to Frank Walsh, April 5, 1932, Cooke Papers, Box 263.

35. Edward O'Neal to Hoover, September 21, 1932, Hoover Papers, Box 410.

36. Samuel F. Rosenman, ed., *Public Papers and Addresses of Franklin D. Roosevelt,* 13 volumes (New York, 1938-1950), I: 733-34.

37. Typescript, Cooke Papers, Box 126.

4: Creation of REA—1935

1. Morris Cooke to Franklin D. Roosevelt, February 20, 1933, Morris L. Cooke Papers, Box 263, Franklin D. Roosevelt Library.

2. Louis Howe to Roosevelt, June 7, 1933, Franklin D. Roosevelt Papers, Box 1, OF 4878, Franklin D. Roosevelt Library.

3. Quoted in Jean Christie, "Morris Llewelyn Cooke: Progressive Engineer." Doctoral dissertation, Columbia University, 1963, p. 208; see also Philip J. Funigiello, *Toward a National Power Policy: The New Deal and the Electric Utility Industry, 1933-1941* (Pittsburg, 1973), p. 130.

4. Cooke to Marvin McIntyre, June 10, 1933, Roosevelt Papers, Box 1, OF 4878.

5. Cooke to McIntyre, July 3, 1933, Roosevelt Papers, Box 1, OF 4878.

6. Bureau of the Census, *Fifteenth Census of the United States, Agriculture,* III, pt. 2, (Washington, 1932): p. 765; ibid., IV, p. 539.

7. David Lilienthal, *TVA, Democracy on the March* (New York, 1944), p. 20; Morris L. Cooke, "The Early Days of the Rural Electrification Idea, 1914-1936," *American Political Science Review,* 42 (June 1948): 444; Harcourt A. Morgan, "Rural Electrification: A Promise to American Life," *Transactions, Third World Power Conference,* 8 (Washington, 1938): 796.

8. Edward Falack, "Operations of Alcorn County Electric Cooperative," typescript, (January 1935), National Agricultural Library.

9. Quote in ibid., p. 4; George Kable, "Rural Electrification in Alcorn County, Mississippi." Paper presented at the American Society of Agricultural Engineers, Athens, Georgia, June 20, 1935, TVA Library; David Lilienthal, "The Tennessee Valley Authority and Farm Electrification." Speech to the American Farm Bureau Federation, December 12, 1934, TVA Library.

10. Quote in Cooke, "Early Days of the Rural Electrification Idea," p. 444; Gordon Clapp, *The TVA: An Approach to the Development of a Region* (Chicago, 1955), p. 105.

11. Thomas G. McGraw, *TVA and the Power Fight, 1933-1939* (Philadelphia, 1971), p. 124; G. D. Munger, "Methods Employed for Financing Equipment and Appliance Purchases." Paper presented at Third World Power Conference, Chicago, September 1936, TVA Library.

12. McGraw, *TVA and the Power Fight,* p. 124.

13. Electric Home and Farm Authority, "Purposes and Program with Recommendations for Expansion," Interdepartmental Report of EHFA, March 15, 1935, TVA Library.

14. David Lilienthal, "Progress in the Electrification of the American Home and Farm." Speech to Chattanooga Chamber of Commerce, September 19, 1934, TVA Library.

15. Christie, "Morris Llewelyn Cooke," p. 215.

16. Quote in John Carmody Interview, Columbia Oral History Collection (COHC), Special Collections Department, Columbia University, p. 402; Clyde T. Ellis, *A Giant Step* (New York, 1966), p. 38.

17. George M. Rommel, "Rural Electrification in the South" typescript (1934), p. 7, TVA Library; quote in *Congressional Record,* 73rd Congress, 2nd Sess., Vol. 78, pt. 1, p. 638.

18. Cooke relates this conversation in "Early Days of the Rural Electrification Idea," pp. 444-45.

19. Ibid.

20. Cooke to Harold Ickes, December 12, 1933, Cooke Papers, Box 263; Cooke to Herman Kahn, July 30, 1959, Cooke Papers, Box 319.

21. Quote in Cooke to Leland Olds, January 29, 1934, Cooke Papers, Box 251; Cooke to Clark E. Mickey, January 29, 1934, Cooke Papers, Box 134.

22. Cooke to Kahn, July 30, 1959.

23. Ibid.

24. Cooke to Ickes, February 13, 1934, Cooke Papers, Box 263.

25. Morris L. Cooke, "National Plan for the Advancement of Rural Electrification Under Federal Leadership and Control with State and Local Cooperation and as a Wholly Public Enterprise" (February 1934), Cooke Papers, Box 230.

26. Ibid.

27. Ibid.

28. Cooke to Ickes, December 12, 1933.

29. McCrory to Rexford G. Tugwell, March 2, 1934, National Archives, Record Group 16.

30. Henry Wallace to Ickes, March 8, 1934, ibid.

31. Cooke to Herman Kahn, September 15, 1959, Cooke Papers, Box 319.

32. Judson King, "Underlying Research to REA Study, 1911-1943," typescript, Box 79, King Papers; L. M. Scott to J. C. B. Ehringhaus, April 14, 1934, J. C. B. Ehringhaus Papers, Box 59, North Carolina State Archives.

33. Mississippi Valley Committee, *Report of the Mississippi Valley Committee of the Public Works Administration* (Washington, 1934); Jean Christie, "The Mississippi Valley Committee: Conservation and Planning in Early New Deal," *The Historian*, 32 (May 1970): 449-69.

34. Christie, "Morris Llewelyn Cooke," pp. 199-200.

35. Cooke to Ickes, October 15, 1934, Box 31, Morris L. Cooke Files, National Archives, Record Group 221; quote in National Resources Board, *A Report on National Planning and Public Works in Relation to Natural Resources and Including Land Use and Water Resources with Findings and Recommendations* (Washington, December 1, 1934), p. 352.

36. Cooke to Ickes, March 22 and 24, 1934, Cooke Papers, Box 263.

37. Christie, "Morris Llewelyn Cooke," p. 218.

38. O'Neal to Wallace, September 19, 1934, Record Group 16.

39. Edward O'Neal Interview, Columbia Oral History Collection, p. 90.

40. David Weaver, "Story of Rural Electrification in North Carolina," *Public Utilities Fortnightly*, 19 (June 1937): 808-15; Thomas F. Ball, "State to Build and Operate its own Rural Electric Lines," *Public Utilities Fortnightly*, 12 (June 1934): 772-80.

41. Albert N. Sanders, "State Regulation of Public Utilities by South Carolina, 1879-1935." Doctoral dissertation, University of North Carolina, 1956, p. 401; J. C. B. Ehringhaus to Roosevelt, November 28, 1934, in David Corbett, ed., *Addresses, Letters and Papers of John Christopher Blucher Ehringhaus, Governor of North Carolina, 1933-1937* (Raleigh, 1950), pp. 389-92; S. H. Hobbs, "A Brief History of Rural Electrification in North Carolina." Presidential address, North Carolina Historical Society, November 1, 1963, North Carolina Room, University of North Carolina Library.

42. For an account of the progress of the relief bill in Congress, see Funigiello, *Toward a National Power Policy*, pp. 136-37. Impetus for a federal plan was aided when Thomas W. Norcross of the Forestry Service finished in March a 235-page report for the National Power Policy Committee, "A New Deal in Rural Electrification. A National Plan." It was written with occasional suggestions only from Cooke who was quite pleased with it. Ickes sent Roosevelt a copy. Cooke to Ickes, March 25, 1935, Box 31, Record Group 221; Ickes to Roosevelt, April 3, 1935, Roosevelt Papers, OF 6.

43. President's Personal File I-P, Press Conference, April 24, 1935, Roosevelt Papers, Box 227.

44. Frank McNinch to Roosevelt, April 27, 1935, Roosevelt Papers, OF 1570. Confusion over the selection of an administrator was real. Lilienthal told Cooke: "After your long gallant fight for rural electrification don't for heaven's sake go away without helping to get it started." Lilienthal to Cooke in Christie, "Morris

Llewelyn Cooke," p. 220; quote in Cooke to Roosevelt, May 3, 1935, Cooke files, Box 30, Record Group 221; FDR Press Releases, No. 201, May 3, 1935, Roosevelt Papers.

5: The Search for an Operations Plan: Triumph of the Cooperative

1. U.S. President, Executive Order 7037, May 11, 1935.
2. H. S. Person, "The Rural Electrification Administration in Perspective," *Agricultural History*, 24 (April 1950): 70-71; quote in Cooke, "The Early Days of the Rural Electrification Idea: 1914-1936," *American Political Science Review*, 42 (June 1948): 446.
3. Person, "Rural Electrification Administration in Perspective," p. 73.
4. Judson King, "Who Will Get the $100,000,000 for Farm Electrification?" *National Popular Government League Bulletin* no. 171 (April 25, 1935), p. 10.
5. *Congressional Record*, 74th Congress, 1st Session, Vol. 79, pt. 7, p. 7574.
6. Ibid., pt. 13, p. 14774.
7. L. K. Richardson, *Wisconsin REA: The Struggle to Extend Electricity to Rural Wisconsin, 1935-1955* (Madison, 1961), p. 24; Thomas McGraw, *TVA and the Power Fight 1933-1939* (Philadelphia, 1971), p. 15.
8. Person, "Rural Electrification Administration in Perspective," p. 73.
9. Ibid.
10. Richardson, *Wisconsin REA*, p. 25; *Business Week* (June 8, 1935), p. 24.
11. Paper presented to Edison Electric Institute, April 25, 1935, Morris L. Cooke Files, Box 30, National Archives, Record Group 221.
12. *Electrical World* (June 8, 1935), p. 58.
13. Ibid. (July 6, 1935), pp. 29-31; *Business Week* (June 8, 1935), p. 23.
14. W. W. Freeman to Cooke, July 24, 1935, Cooke Files, Box 16.
15. Kenneth Trombley, *The Life and Times of a Happy Liberal, Morris Llewelyn Cooke* (New York, 1954), pp. 148-49; Marquis Childs, *Farmer Takes a Hand*, (New York, 1953), pp. 58-59; Person, "Rural Electrification Administration in Perspective," p. 74.
16. Cooke to W. W. Freeman, July 31, 1935, Cooke files, Box 16.
17. John W. McCormack Interview, July 18, 1969, Sam Rayburn Oral History Collection, Sam Rayburn Library. Public power enthusiasts had long insisted that the holding company with its alleged demand for profit instead of service was responsible for the lack of rural electrification.
18. Philip Funigiello, *Toward a National Power Policy*, (Pittsburgh, 1973), p. 98.
19. Ibid., p. 113.
20. William Leuchtenburg, *Franklin D. Roosevelt and the New Deal* (New York, 1963), pp. 156-57.
21. Quote in Paper presented to Edison Electric Institute. Richardson, *Wisconsin REA*, p. 25.
22. *Rural Electrification News*, 1 (October 1935): 21.
23. *Transcript of REA General Staff Conference*, February 1-5, 1937, John M. Carmody Papers, Box 95, Franklin D. Roosevelt Library.
24. Walter Pierce to Cooke, August 13, 1935, Cooke Files, Box 28.

25. Julia E. Johnson, ed., *The Reference Shelf, Government Ownership of Electric Utilities* (New York, 1936), p. 235.

26. Murray Lincoln, *Vice-President in Charge of Revolution* (New York, 1960), p. 113.

27. John Carmody Interview, Columbia Oral History Collection (COHC), Special Collections Department, Columbia University, p. 372.

28. "Conference of Co-op Representatives and REA," typescript, (June 6, 1935), p. 23, National Agricultural Library.

29. Cooke to C. W. Warburton, June 10, 1935, Cooke Files, Box 35.

30. Cooke to Boyd Fisher, July 2, 1935, *ibid*, Box 14.

31. REA Public Release no. 25, August 10, 1935, Judson King Papers, Box 79, Library of Congress, Manuscript Division; Richardson, *Wisconsin REA*, p. 28.

32. Richardson, *Wisconsin REA*, p. 27.

33. REA News Releases, November 4 and 6, 1935, Cooke Files, Box 31.

34. Quote in *Rural Electrification News*, 1 (December 1935): 2; *New York Times*, December 29, 1935, pt. III, p. 10.

35. Carmody, Interview, COHC, pp. 376-77.

36. Cooke to Roosevelt, January 1, 1936, Roosevelt Papers, Box 1, OF 1570; Memorandum, September 12, 1935, Roosevelt Papers, Box 1, OF 1570; Roosevelt to Director of the Bureau of the Budget, September 12, 1935, Roosevelt Papers, Box 2, OF 79; House of Representatives, Committee on Interstate and Foreign Commerce, 74th Congress, 2nd Session, *Hearings, A Bill to Provide for Rural Electrification and for other Purposes* (Washington, March 12-14, 1936), pp. 22-23; *Transcript of the REA General Staff Conference*, February 1-5, 1937, p. 10, Carmody Papers, Box 95; *New York Times*, February 16, 1936, p. 33; Albert B. Gerber, "Federal Administration of Rural Electrification." Masters thesis, George Washington University, 1942, p. 23.

37. "Report of the Rural Electrification Administration to the National Emergency Council" (October 14, 1935, Washington, D.C.), p. 11, typescript, National Agricultural Library.

38. Norris to Cooke, October 24, 1935, George W. Norris Papers, Tray 71, Box 9, Manuscript Division, Library of Congress.

39. REA News Release, October 25, 1935; Cooke to Roosevelt, November 12, 1935, Roosevelt Papers, Box 1, OF 1570.

40. Cooke to Norris, November 14, 1935, reprinted in *Rural Electrification News*, 1 (November 1935): 8-9.

6: REA Made Permanent—1936

1. *Rural Electrification News*, 1 (December 1935): 15-16.

2. Ibid., p. 16.

3. E. H. Lightner to Norris, October 26, 1935, George W. Norris Papers, Tray 72, Box 9, Manuscript Division, Library of Congress.

4. *Rural Electrification News*, 2 (March, 1936): 3; *New York Times*, February 16, 1936, p. 33.

5. Clyde T. Ellis, *A Giant Step*, (New York, 1966), p. 46.

6. *Rural Electrification News*, 1 (March 1936): 3-4.

7. "Analysis: Norris-Rayburn Farm Electrification Bill," submitted by Hudson Reed, typescript (February 14, 1936); *Baltimore Sun*, February 18, 1936; Cooke to John P. Robertson, March 2, 1936—all in Morris L. Cooke Files, Box 30, Record Group 221, National Archives, Manuscript Division.

8. Harry Truman to Norris, February 24, 1936. Norris Papers, Tray 71, Box 9.

9. Roosevelt to Secretary of the Treasury, March 5, 1936, Franklin D. Roosevelt Papers, Box 1, OF 1570; Cooke to Sam Rayburn, March 16, 1936, Cooke Files, Box 30; *Washington Post*, February 27, 1936; Cooke to Norris, February 28, 1936, Norris Papers, Tray 71, Box 9; *Congressional Record*, 74th Congress, 2nd Sess., Vol. 80, pt. 3, pp. 2750-59, 2819-833. One writer asserts that Norris and Jesse Jones had agreed on the lower figure earlier and the reduction in the funding was staged as a concession to the opposition. Evidence is not conclusive, but it would be un-usual for Norris to plan his strategy without Cooke. See Mark Stauter, "The Rural Electrification Administration, 1935-1945: A New Deal Case Study." Doctoral dissertation, Duke University, 1973, pp. 59-62.

10. Norris agreed to the shorter amortization specifically because distribution poles had a twenty-five year lifespan, and it was thought best not to let the loans exceed that period. See W. E. Herring to Cooke, February 24, 1936, Norris Papers, Tray 71, Box 9; *Congressional Record*, 74th Congress, 2nd Sess., Vol. 80, pt. 3, pp. 2819-833; *New York Times*, March 5, 1936, p. 5.

11. For a discussion of the Holding Company Act, see Philip Funigiello, *Toward a National Power Policy* (Pittsburgh, 1973), pp. 67-97; John M. Carmody Interview, Columbia Oral History Collection (COHC), Special Collections, Columbia University, p. 402.

12. Deward C. Brown, "The Sam Rayburn Papers: A Preliminary Investigation," *The American Archivist*, 31 (July-October 1972): 331-36.

13. Rayburn to Thurman Kerr, February 12, 1949, Samuel T. Rayburn Papers, Sam Rayburn Library.

14. U.S. House of Representatives, Committee on Interstate and Foreign Com-merce, 74th Congress, 2nd Sess., *Hearings, A Bill to Provide for Rural Electrifica-tion and for other Purposes*, pp. 60-61.

15. Ibid., pp. 55-56.

16. Ibid., pp. 96-103; H. S. Person, "The Rural Electrification Administration in Perspective," *Agricultural History*, 24 (April 1950): 75.

17. W. R. Poage to D. Clayton Brown, Interview, July 16, 1969; *Birmingham News*, March 17, 1936; *Knoxville News-Sentinel*, March 18, 1936; Virginius Dabney, *Below the Potomac* (New York, 1942), pp. 306-08.

18. Ibid.

19. *Congressional Record*, 74th Congress, 2nd Sess., Vol. 80, pt. 5, pp. 5273-318.

20. Ibid., pp. 5275-318.

21. Ibid., p. 5315.

22. Ibid.

23. George Norris, *Fighting Liberal* (New York, 1945), p. 321.

24. George Norris, "Rural Electrification," in *Transcript of the First Meeting of the NRECA,* 1944, p. 142, Files of Administrator William Neal, 1938-1948, Record Group 221.

25. Ibid., p. 144.

26. Memorandum, E. J. Coil to Cooke, November 11, 1935, Cooke Files, Box 6; D. S. Weaver to Cooke, November 16, 1935, Roosevelt Papers, Box 5, OF 1570; Cooke to Roosevelt, November 20, 1935, ibid.; Committee on Interstate and Foreign Commerce, *Hearings, a Bill to Provide Rural Electrification,* pp. 55-56.

27. Quote in Norris to John O'Connor, February 1, 1936, Norris Papers, Tray 71, Box 6; Norris to Walter Pierce, March 5, 1936, Norris Papers, Tray 71, Box 9.

28. Cooke to Norris, May 6, 1936, Cooke Files, Box 30.

29. Cooke to Roosevelt, May 7, 1936, Roosevelt Papers, Box 1, OF 1570; quote in Norris to Walter Pierce, March 6, 1936, Norris Papers, Tray 71, Box 9.

30. U.S. House of Representatives, 74th Congress, 2nd Sess., *Rural Electrification Act of 1936,* Conference Report no. 2644; Ellis, *A Giant Step,* pp. 48-49.

31. U.S. House of Representatives, 74th Congress, 2nd Sess., *Rural Electrification Act of 1936,* Conference Report no. 2219.

32. *Rural Electrification News,* 1 (June 1936): 3.

33. Ibid., p. 5.

34. *Moody's Public Utility Manual, 1974* (New York), Special Features Section, p. a6; *Rural Electrification News,* 10 (October 1944): 4.

35. Cooke to John Carmody, April 13, 1939, Cooke Papers, Box 147.

7: Electricity Comes to the Farm

1. *Rural Electrification News,* 1 (August 1936): 23; quote in Mark C. Stauter, "The Rural Electrification Administration, 1935-1945, A New Deal Case Study, Doctoral dissertation, Duke University, 1973, p. 74.

2. Ibid., p. 82.

3. John M. Carmody Interview, Columbia Oral History Collection (COHC), Special Collections, Columbia University, p. 482.

4. Jilson McCullough to Sam Rayburn, July 19, and December 9, 1937, Samuel T. Rayburn Papers, Sam Rayburn Library; Murray Lincoln, *Vice-President in Charge of Revolution* (New York, 1960) pp. 133-37.

5. Cooke to Edward O'Neal, June 23, 1936, Morris L. Cooke Files, Box 26, National Archives, Record Group 221; *Rural Electrification News,* 1 (July 1936): 11-13; ibid., 4 (August 1939): 14; James B. Richey, first manager of the Fannin County Cooperative, Bonham, Texas, to D. Clayton Brown, Interview, December 23, 1968.

6. For a discussion of the founding of a co-op, see Delbert E. Taylor, "History of the Fannin County Rural Electric Cooperative." Master's thesis, East Texas State University, 1964.

7. Harry Slattery, "Rural Electrification for the Millions," typescript, [n.d.], Judson King Papers, Box 79, Library of Congress, Manuscript Division.

8. Ibid.; *Rural Electrification News,* 4 (January 1939): 3-6.

9. Ibid.

10. W. Fred Jordon, *The Arkansas Plan*, REA Interdepartmental Pamphlet (June 1940), National Agricultural Library. For the origins of this plan, see D. Clayton Brown, "Hen Eggs to Kilowatts: Arkansas Rural Electrification," *Red River Valley Historical Review*, 3 (Winter 1978): 119-25.

11. REA, *Annual Report, 1940* (Washington 1940), p. 9.

12. Carmody to Cochran, January 26, 1955, John M. Carmody Papers, Box 83, Franklin D. Roosevelt Library.

13. Harry Slattery, *Rural America Lights Up* (Washington, 1940), p. 121; *Rural Electrification News*, 3 (November 1937): 3-4; ibid., 1 (June 1936): 19.

14. Cooke to Ehringhaus, October 27, 1936, J. C. B. Ehringhaus Papers, Box 59, North Carolina State Archives.

15. Cooke to Paul Appleby, September 5, 1935, Cooke Files, Box 7.

16. James H. Rogers to Cooke, November 13, 1935, Cooke Files, Box 7; "Notes on REA," General Correspondence, AL-W 1936, King Papers, Folder no. 18; REA Memorandum, Carmody to Clay Cochran, January 26, 1955, Carmody Papers, Box 83; Carmody Interview, COHC, p. 454.

17. Carmody to Judson King, July 26, 1944, Carmody Papers, Box 83.

18. E. E. Taylor, Field Report, Bessier Rural Electric Membership Corporation, Records Relating to Spite Line Activity, 1940-1941, National Archives, Record Group 221; Joel E. Lennon to Sam Rayburn, July 16, 1937, Rayburn Papers; quote in Carmody to Cochran, January 26, 1955, Carmody Papers, Box 83.

19. W. W. Freeman to Cooke, July 24, 1935, Cooke Files, Box 16.

20. Marquis Childs, *The Farmer Takes a Hand* (New York, 1953), p. 77.

21. *Moody's Public Utility Manual* (New York, 1954), p. A13.

22. *Congressional Record*, 79th Congress, 2nd Session, Vol. 92, pt. 2, p. 2114; Twentieth Century Fund, *Electric Power and Government Policy* (New York, 1948), p. 447.

23. Twentieth Century Fund, *Electric Power and Government Policy*, p. 465.

24. Quoted in *Rural Electrification News*, 4 (August 1939): 14.

25. *REA Co-op Message*, Bulletin (July 1938), Greenville, Texas, Rayburn Papers.

8: The Slattery Incident—1939-1944

1. George Norris to Roosevelt, June 15, 1939, Franklin D. Roosevelt Papers, Box 1, OF 1570, Franklin D. Roosevelt Library. Reasons for the transfer are unknown. See Philip Funigiello, *Toward a National Power Policy* (Pittsburgh, 1973), pp. 165-66; Clyde T. Ellis, *A Giant Step* (New York, 1966), pp. 62-63; Carmody to Cooke, July 1, 1939, Morris L. Cooke Papers, Box 147, Franklin D. Roosevelt Library.

2. Harry Slattery, "From Roosevelt to Roosevelt: A Study of Forty Years in Washington, TR to FDR, 1905 to 1945." Unpublished autobiography, p. 247, Harry Slattery Papers, Duke University Library.

3. Quoted in Mark C. Stauter, "The Rural Electrification Administration, 1935-1945, A New Deal Case Study." Doctoral dissertation, Duke University, 1973, p. 142.

4. REA, *1953 Annual Statistical Report* (Washington, 1954), p. xv.

5. Slattery, "From Roosevelt to Roosevelt," pp. 246-47.

6. *New York Times,* April 30, 1967, p. 87; Dean Albertson, *Roosevelt's Farmer: Claude R. Wickard in the New Deal* (New York, 1955), pp. 6-18, 132-33, 154-55, 170, 177.

7. Slattery, "From Roosevelt to Roosevelt," pp. 255-58.

8. Ibid., pp. 259-60.

9. Samuel B. Bledsoe Interview, Columbia Oral History Collection (COHC), Special Collections Department, Columbia University, p. 399.

10. Slattery, Confidential Memorandum, February 20, 1941, Harry Slattery Papers, Box 109, Duke University, Manuscript Collection.

11. Ibid., February 8, 10, 1941, Slattery Papers, Box 109; March 3, 1941, Slattery Papers, Box 65.

12. Boyd Fisher, "History of Controversy in REA" typescript, John M. Carmody Papers, Box 85, Franklin D. Roosevelt Library.

13. Ibid., p. 52.

14. Ibid., p. 53.

15. Bledsoe Interview, COHC, pp. 400-01.

16. Ibid., p. 398.

17. David Lilienthal, *The Journals of David E. Lilienthal, The TVA Years,* I (New York, 1964), p. 280.

18. Memorandum, King to Robert Craig, June 20, 1941, Judson King Papers, Box 78, Library of Congress, Manuscript Division.

19. *Business Week* (January 24, 1942), p. 60.

20. *Labor, a National Weekly* (April 22, 1944), Slattery Papers, Boxes 141-42; *Electrical World,* 121 (April 22, 1944): 241; Stauter, "The Rural Electrification Administration," p. 187.

21. REA, *1953 Annual Statistical Report* (Washington, 1954), p. xv.

22. Slattery, "From Roosevelt to Roosevelt," p. 271; Lilienthal, *The TVA Years,* p. 404.

23. *Electrical World,* 118 (August 8, 1942): 859.

24. Charles M. Curfman to Rayburn, March 26, 1942, Samuel T. Rayburn Papers, Sam Rayburn Library. For further discussion of the WPB freeze, see H. S. Person, "The Rural Electrification Administration in Perspective," *Agricultural History,* 24 (April 1950): 79-80.

25. Slattery, "From Roosevelt to Roosevelt," p. 271; Ellis, *A Giant Step,* p. 70.

26. H. H. Sears to Rayburn, January 28, 1942, Rayburn Papers.

27. Lilienthal, *The TVA Years,* pp. 404-05.

28. Charles Curfman to Rayburn, February 5, 1942, Rayburn Papers; *Dallas Morning News,* February 1, 1942; *Congressional Record,* 77th Congress, 1st Sess., Vol. 87, pt. 9, p. 9315; *Electrical World,* 117 (February 14, 1942): 1; quoted in Stauter, "The Rural Electrification Administration," p. 201.

29. Stauter, "The Rural Electrification Administration," pp. 195-96; Ellis, *A Giant Step,* p. 71.

30. *Dallas Morning News,* February 1, 1942; *Congressional Record,* 77th Congress, 2nd Sess., Vol. 88, pt. 2, p. 2004; *Electrical World,* 117 (February 14, 1942): 3.

31. Quoted in Donald H. Cooper, "The Advance of Rural Electrification, 1940-1945," typescript, (1972), p. 2, NRECA Library, Washington, D.C.

32. Ellis, *A Giant Step,* pp. 71-72. Brief mention was made of a national association in Cooke's correspondence with Edward O'Neal. See O'Neal to Cooke, March 18, 1936, Morris L. Cooke Files, Box 25, National Archives, Record Group 221.

33. United States Senate, Subcommittee of the Committee on Agriculture and Forestry, 78th Congress, 1st Sess., *Hearings on Senate Resolution 197,* pt. 5, p. 1906.

34. Clyde T. Ellis to D. Clayton Brown, Interview, August 5, 1969, Washington, D.C.; quote in Ellis, *A Giant Step,* p. 29.

35. Ellis to Brown, Interview.

36. Ellis, *A Giant Step,* pp. 73-74.

37. Ibid., p. 80; *Electrical World,* 119 (March 23, 1943): 1043; REA, *1943 Annual Report* (Washington, 1943), p. 3; Person, "The Rural Electrification Administration in Perspective," p. 80; *New York Times,* July 2, 1943, p. 5.

38. Robert Craig to Sam Rayburn, January 18, 1943, Rayburn Papers; Rayburn to Luke Doyle, January 25, 1943, Rayburn Papers; *Electrical World,* 119 (January 9, 1943): 72; Albertson, *Roosevelt's Farmer,* p. 358.

39. "Chronology of NRECA-REA Controversy," typescript, King Papers, Box 79; *Rural Electrification News,* 3 (June 1938): 7-8.

40. Marquis Childs, *The Farmer Takes a Hand* (New York, 1953), p. 97; Ellis, *A Giant Step,* pp. 75-77; Senate, *Hearings on Senate Resolution 197,* pt. 2, pp. 421-28.

41. John Asher to Slattery, Senate, *Hearings on Senate Resolution 197,* June 5, 1943, p. 428.

42. Senate, *Hearings on Senate Resolution 197,* pt. 2, pp. 577-78.

43. Ibid., pp. 215, 276, 307.

44. Officers of the NRECA to Franklin D. Roosevelt, May 8, 1943, Slattery Papers, Box 73; NRECA to Members, June 5 and June 7, 1943, Slattery Papers, Box 74; John George to Northern Idaho Rural Electric Rehabilitation Association, June 15, 1943, Slattery Papers, Box 74; Judson King, "The Inside of the Cup," *The National Popular Government League Bulletin No. 208* (September 10, 1943), pp. 3-4; *St. Louis Post-Dispatch,* May 20, 1943.

45. Senate, *Hearings on Senate Resolution No. 197,* pt. 5, pp. 2110-120; Slattery to Alabama REA Cooperatives, February 27, 1943, Slattery Papers, Box 72; Slattery to Wickard, February 27, 1943, Slattery Papers, Box 72.

46. Edgar A. Brown to Slattery, June 8, 1943; N. E. Hubel to Louis R. Glavis, June 4, 1943; Slattery to Wickard, June 15, 1943; Norris to Ellis, June 25, 1943—all in Slattery Papers, Box 74; Judson King, "What is the True Origin of the NRECA?" *Public Utilities Fortnightly* in Senate, *Hearings on Senate Resolution No. 197,* pt. 5, p. 1919.

47. Cooke to Slattery, July 29, 1943, Slattery Papers, Box 75.

48. Pinchot to Slattery, July 30, 1943, ibid.

49. Quote in *Doying—Washington,* typescript (April 25, 1941), Slattery Papers, Box 109; *New York Times,* May 25, 1951, p. 27; *Rural Electrification News,* 3 (December 1937): 13-14.

50. Truett Bailey, Chairman, Board of Directors, Brazos River Co-op, to D. Clayton Brown, Interview, August 23, 1973.

51. Elliot Roosevelt to Steven Early, March 1, 1941, reprinted in Senate, *Hearings on Senate Resolution No. 197*, pt. 2, pp. 396-400; Bailey to Brown, Interview; *St. Louis Post-Dispatch*, May 23, 1944.

52. Quote in Robert S. Broderick to Slattery, June 7, 1943; Slattery to Rayburn, June 3, 1943; Memorandum, Harry Slattery, June 11, 1943—all in Slattery Papers, Box 74.

53. Wickard to Marvin McIntyre, June 25, 1943, Roosevelt Papers, OF 1570; Slattery, "From Roosevelt to Roosevelt," p. 293; *St. Louis Post-Dispatch*, December 10, 1943; quote in ibid., October 6, 1944.

54. J. W. Fullbright to Roosevelt, June 9, 1943; Manson Morris to Roosevelt, May 14, 1943; Ed Rodden to "Missy," March 29, 1940; Wickard to Marvin McIntyre, May 27, 1943, June 6, 1943; Poage to Roosevelt, July 10, 1943—all in Roosevelt Papers, OF 1570; *Business Week* (July 10, 1943), pp. 27-28; Jonathan Daniels, *White House Witness, 1942-1945* (New York, 1975), p. 172.

55. Ibid.

56. Slattery, "From Roosevelt to Roosevelt," p. 294.

57. Quoted in Stauter, "The Rural Electrification Administration," pp. 242-43.

58. Senate, *Hearings on Senate Resolutions No. 197*, pt. 1, p. 1; Slattery, "From Roosevelt to Roosevelt," p. 298.

59. For the official transcript, see Senate, *Hearings on Senate Resolution No. 197*, pts. 1-5; "Rural Electrification," Files of the American Farm Bureau Headquarters, Policy Statements, to D. Clayton Brown, March 28, 1979.

60. Daniels, *White House Witness*, p. 207.

61. Quote in ibid., p. 218.

62. *Congressional Record*, 78th Congress, 2nd Sess., Vol. 90, pt. 5, pp. 6591-594.

63. *New York Times*, December 12, 1944, p. 1; *St. Louis Post-Dispatch*, June 19 and September 22, 1944.

64. Slattery to Rayburn, December 14, 1943, Rayburn Papers.

65. *Congressional Record*, 78th Congress, 2nd Sess., Vol. 90, pt. 2, pp. 2283-287.

66. Ibid., pt. 3, pp. 3834-851; *REA News* (September 1944), p. 4; Person, "The Rural Electrification Administration in Perspective," p. 80.

67. *Ibid*. REA, *1953 Annual Statistical Report*, p. xv.

9: The Postwar Fight for Energy in the South

1. Quote in United States Congress, House Committee on Flood Control, *Flood Control on the Mississippi River: Hearings*, Part 3, April 28 to May 2, 1930, 71st Congress, 2nd Sess., 746-47. Portions of this chapter were published in *Southwestern Historical Quarterly*, 73 (October 1974): 140-54.

2. *Sherman* (Tex.) *Daily Democrat*, June 19, 1938; *Dallas Times Herald*, August 7, 1943; United States Department of the Interior, *Annual Report of the Secretary of the Interior: Fiscal Year Ended June 30, 1945* (Washington, 1945), p. 68.

3. *Congressional Record,* 77th Congress, 1st Sess., Vol. 87, pt. 12, Appendix, pp. 2621-622; Clyde T. Ellis, *A Giant Step* (New York, 1966), p. 30.

4. Ibid., p. 32.

5. United States Department of the Interior, *Annual Report of the Secretary of the Interior for the Fiscal Year Ended June 30, 1944* (Washington, 1944), p. 64.

6. J. W. Madden, to Rayburn, February 22, 1937, Samuel T. Rayburn Papers, Sam Rayburn Library; J. E. Smitherman to Rayburn, July 12, 1937, Rayburn Papers.

7. Charles Curfman to Rayburn, January 25, 1943, Rayburn Papers; Rayburn to Denison *Herald,* October 1, 1943, Rayburn Papers; Douglas G. Wright, First Administrator of the SPA, to D. Clayton Brown, Interview, June 11, 1969; Southwestern Power Administration, *History of the SPA* (Tulsa, 1952), pp. 25-28.

8. L. H. Wentz to Rayburn, April 9 and April 24, 1942, Rayburn Papers; Wright to Brown, Interviews, June 11, 1969 and February 5, 1972.

9. Memorandum, Secretary of the Interior to Franklin D. Roosevelt, August 20, 1935, Franklin D. Roosevelt Papers, Box 1, OF 1570, Franklin D. Roosevelt Library; National Resources Committee, "Recommendations and Conclusions Concerning the Pensacola Project of Oklahoma," August 15, 1935, typescript, Morris L. Cooke Files, Box 31, Record Group 221, National Archives; W. E. Herring to Morris L. Cooke, May 14, 1934, Morris L. Cooke Papers, Box 263, Franklin D. Roosevelt Library; Alvin J. Wirtz to Rayburn, June 7, 12, 1943, Rayburn Papers; Wright to Brown, Interviews.

10. This fight within the executive branch is discussed in Harold Ickes, *The Lowering Clouds, 1939-1941,* Vol. III of *The Secret Diary of Harold L. Ickes* (New York, 1954), pp. 22, 28, 491, 504, 578-87, 619. See also Wright to Brown, Interviews; Clyde T. Ellis to D. Clayton Brown, Interviews, August 5-6, 1969. For background to the Ickes-Olds struggle, consult Philip J. Funigiello, "Kilowatts for Defense: The New Deal and the Coming of the Second World War," *Journal of American History,* 56 (December 1969): 604-20.

11. Wirtz to Rayburn, June 12, 1943, Rayburn Papers; Wright to Brown, Interviews; Ellis to Brown, Interviews; *New York Times,* February 16, 1944.

12. Wirtz to Rayburn, June 12, 1943, Rayburn Papers.

13. Rayburn to Roosevelt, July 23, 1943, ibid.

14. Rayburn to Byrnes, July 26, 1943, ibid.

15. Roosevelt to Rayburn, July 30, 1943, ibid. The Grand River dam was temporarily placed in the hands of Secretary Ickes, and for a short time it was part of the SPA. In 1946, however, the structure was returned to the state of Oklahoma. Roosevelt placed Grand River in the Interior Department in order to simplify the bureaucratic operation of all dams in the Southwest that supplied defense industries. Until the president granted Rayburn's wish, it appeared that the Grand River project would stay in the FWA. See *Co-op Power,* 2 (December 1945): 12-13.

16. *Denison* (Tex.) *Herald,* July 31, 1943; *Sherman* (Tex.) *Daily Democrat,* August 10, 1943.

17. Ickes, *Lowering Clouds,* pp. 580, 585-88; David E. Lilienthal, *The Journals of David E. Lilienthal, The TVA Years* (New York, 1964), pp. 664, 666-68. The Rayburn Papers document the Roosevelt-Rayburn relationship, although sparsely. See Deward C. Brown, "The Sam Rayburn Papers: A Preliminary Investigation,"

The American Archivist, 35 (July-October 1972): 331-336. For correspondence pertaining to Rayburn's role in the 1932 Democratic National Convention, see William G. McAdoo to Rayburn, September 20, 1938, February 20 and April 28, 1939, Rayburn Papers; Rayburn to McAdoo, March 3, 1939, Rayburn Papers.

18. Roy F. Hall to Rayburn, March 29, 1943, Rayburn Papers.

19. F. W. Rossom to Rayburn, March 12, 1946, ibid.

20. M. L. Griffin to Rayburn, March 15, 1946, ibid.

21. Southwestern Power Administration, *History of the SPA,* pp. 8-9; *New York Times,* December 10, 1944, p. 48.

22. *New York Times,* November 25, 1944, p. 14; *Congressional Record,* 78th Congress, 2nd Sess., Vol. 90, pt. 6, pp. 8227-255.

23. Deward C. Brown, "Rural Electrification in the South, 1920-1955." Doctoral dissertation, University of California, Los Angeles, 1970, p. 237.

24. Marquis Childs, *The Farmer Takes a Hand* (New York, 1953) pp. 174-75.

25. Quote in *Rural Electrification* 3 (November 1955): 24; Rayburn to Thurman Kerr, February 12, 1949, Rayburn Papers.

26. *Business Week* (June 12, 1943), p. 38; William A. Settle, Jr., *The Dawning, A New Day for the Southwest: A History of the Tulsa District Corps of Engineers, 1939-1971* (Tulsa, Okla., 1975), pp. 50-51.

27. Quote in Wright to Rayburn, February 21, 1947; James Buster to Rayburn, January 29, 1941; Claude Easterly to Rayburn, October 1, 1943—all in Rayburn Papers; *Denison* (Tex.) *Herald,* July 31, 1943.

28. *New Republic* (July 28, 1947), pp. 9-10; *Saturday Evening Post* (January 19, 1946), pp. 22-23; *Congressional Digest,* 29 (January 1950): 26-31; Wesley C. Clark, "Proposed 'Valley Authority' Legislation," *The American Political Science Review,* 40 (February 1946): 62-70.

29. Southwestern Power Administration, *History of the SPA,* pp. 2-3.

30. "Cheap Power at Issue in Southwest," *Co-op Power,* 2 (December 1945), pp. 12-13; Wright to D. Clayton Brown, Interview, February 5, 1972; *Business Week* (December 25, 1948), pp. 22-23.

31. *Rural Electrification,* 4 (July 1946): 30.

32. *Congressional Record,* 79th Congress, 2nd Sess., Vol. 92, pt. 4, pp. 5131-132.

33. Ibid., pt. 2, p. 2115; *New York Times,* March 12, 1946, p. 26.

34. U.S. House of Representatives, Subcommittee on Appropriations, *Interior Department Appropriations for 1947: Hearings . . .: Part 3: Southwestern Power Administration,* 79th Congress, 2nd Sess., pp. 136-205; *Rural Electrification,* 4 (July 1946): 7, 30; Elmo Thomas to Russel Smith, June 5, 1946; E. W. Higgins to Rayburn, May 8, 1946; James G. Patton to Arthur Cross, June 6, 1946—all in Rayburn Papers; Childs, *Farmer Takes a Hand,* p. 100.

35. Frank M. Wilkes, "It CAN Happen Here—And It HAS!" *Electrical South,* 33 (July 1953): 18 (quotation), 19-21; Rayburn to Higgins, June 10, 1946, Rayburn Papers; Thomas to Smith, June 5, 1946, Rayburn Papers; *Tulsa Tribune,* February 8, 1947; "Summary of Utility Groups' Offer to Distribute Federal HydroPower," *Public Utilities Fortnightly,* 38 (October 24, 1946): 564-65; *Business Week* (April 29, 1950), p. 21.

36. Quoted in *Rural Electrification*, 10 (August 1952): 26; Rayburn to Truman, January 10, 1947, Rayburn Papers.

37. Wright to D. Clayton Brown, Interview, February 5, 1972; Rayburn to Truman, January 10, 1947, Rayburn Papers; Southwestern Power Administration, *History of the SPA*, 1 (second quotation), 1-2; U.S. House of Representatives, Subcommittee on Appropriations, *Interior Department Appropriations for 1949: Hearings . . .: Part 3, Southwestern Power Administration*, 80th Congress, 2nd Sess., pp. 89-90.

38. U.S. House of Representatives, Subcommittee on Appropriations, *Interior Department Appropriations for 1947: Hearings . . .. Part 3: Bonneville Power Administration*, 79th Congress, 2nd Sess., pp. 360-63, 394-98.

39. Southwestern Power Administration, *Missions, Resources and Program Objectives* (February 1969), pp. 1-2.

40. Ibid. The figures of the distribution of SPA energy were: REA co-ops, 53 percent; municipalities, 12 percent; defense industry installations, 13 percent; and private utility companies, 22 percent.

41. Southwestern Power Administration, *History of the SPA*, p. 3.

42. Department of the Interior, *Annual Report, 1950* (Washington, 1951), p. 401; *Chattanooga Times*, February 2, 1950.

43. *Henderson* (N.C.) *Dispatch*, January 6, 1942; *Nashville Banner*, May 14, 1952; U.S. House of Representatives, Subcommittee on Appropriations, *Interior Department Appropriations for 1952*, 82d Congress, 1st Sess., pp. 85-117.

44. *New York Times*, March 19, 1950, III, p. 1; *U.S. News and World Report* (June 6, 1952), p. 56; Childs, *Farmer Takes a Hand*, pp. 215-31.

10: America Achieves Rural Electrification

1. "Rural Electrification: A Postwar Market Forecast," *Country Gentleman*, 114 (Philadelphia, 1944), reprint in National Agricultural Library.

2. Harry Slattery to Morris Cooke, October 24, 1939, Morris L. Cooke Papers, Box 147, Franklin D. Roosevelt Library; Roosevelt to John Rankin, July 14, 1941, Franklin D. Roosevelt Papers, Box 3, OF 1570, Franklin D. Roosevelt Library; quote in REA Postwar Planning Committee, *Rural Electrification After the War, A Preliminary Report* (St. Louis, 1944), p. 2.

3. Rural Electrification Administration, *Annual Statistical Report, 1953* (Washington, 1953), p. xv.

4. Marquis Childs, *The Farmer Takes a Hand* (New York, 1953), p. 110.

5. *Moody's Public Utility Manual* (New York, 1954), p. A-13.

6. Ibid.

7. Ibid.; REA, *Statistical Annual Report, Rural Electric Borrowers*, REA Bulletin 1-1 (October 1973), p. 9.

8. Twentieth Century Fund, *Electric Power and Government Policy* (New York, 1948), p. 477.

9. Ibid., pp. 457-62, 476-78.

10. National Rural Electric Cooperative Association, *Rural Electric Fact Book* (Washington, 1965), pp. 118-26.

11. Joe F. Davis, "Economic Studies of Farm Electrification," *Agricultural Engineering,* 31 (November 1950): 565-68.

12. Louisan Mamer, "Electricity Pays Its Own Way in the Rural Home," typescript (March 11, 1952), National Agricultural Library, p. 10; *Rural Electrification News,* 4 (March 1939): 22; quote in ibid., p. 14.

13. Mamer, "Electricity Pays its Own Way," p. 3; quote in *Rural Electrification News,* 4 (May 1939): 11.

14. REA Interbureau Coordinating Committee, "Present Uses of Electricity in Rural Areas," typescript (1941), p. 1, Mercer Green Johnston Papers, Manuscript Division, Library of Congress; Frederick Mott and Milton I. Roemer, *Rural Health and Medical Care* (New York, 1948), pp. 34, 91, 111.

15. *Rural Electrification News,* 3 (October 1937): 4.

16. Ibid., p. 3.

17. Mollie Ray Carroll, "General Conclusions from a Survey of the Moral and Cultural Effects of the Inside Bathroom and Refrigerator," typescript (1935), Carroll to D. Clayton Brown.

18. Victor Ehler and Ernest Steel, *Municipal and Rural Sanitation* (New York, 1943) p. 313; quote in Rello H. Britten, "The Relation Between Housing and Health," *Public Health Reports,* 44 (November 2, 1934): 1308.

19. James F. Evans, *Prairie Farmer and WLS: The Burridge D. Butler Years* (Urbana, Ill., 1969), pp. 153-75; Edward S. de Brunner, *Radio and the Farmer,* Pamphlet, Radio Institute of the Audible Arts (New York, 1936); C. W. Jackson, "A Study of the Value of the Texas Farm and Home Radio Program in Twenty Northeast Texas Counties," typescript (May 1946), National Agricultural Library; *Rural Electrification News,* 8 (June 1943): 10, 22.

20. Ibid., 4 (October 1938): 5; *Rural Electrification Exchange,* 1 (April 1938): 33.

21. Quote in Thomas D. Clark, *The Emerging South* (New York, 1961), pp. 90-91; USDA Bureau of Agricultural Economics, "Electricity on Farms in the Upper Piedmont of Georgia," *Georgia Experiment Station Bulletin no. 263* (June 1950), p. 34.

22. *Progressive Farmer* (April 1956), p. 46.

23. *Rural Electrification News,* 4 (March 1939): 14.

11: Ideology of Rural Electrification

1. Examples of this attitude are numerous. David Cushman Coyle, *Electric Power on the Farm,* REA special pamphlet (Washington 1936); George Norris, *Fighting Liberal* (New York, 1945), pp. 318-19; O. M. Rau, "Rural Electric Lines to Curb Migration from Farms," *Public Utilities Fortnightly,* 16 (August 15, 1935): 200-09; Clyde T. Ellis, *A Giant Step* (New York, 1966), pp. 176-78; E. R. Eastman, *These Changing Times* (New York, 1927), pp. 45-54.

2. Connections between business and progressive reforms are set forth in Robert Wiebe, *Businessmen and Reform: A Study of the Progressive Movement* (Cambridge, 1962); Gabriel Kolko, *The Triumph of Conservatism: A Reinterpretation of American History, 1900-1916* (New York, 1963).

3. Jean Christie, "Giant Power: A Progressive Proposal of the Nineteen-Twenties," *Pennsylvania Magazine of History*, 96 (October 1972): 489.

4. Morris L. Cooke to Hoover, October 20, 1931, Hoover Papers, Box 394, Herbert Hoover Library.

5. For varied interpretations of Hoover, see Martin Fausold and George T. Mazuzan, eds., *The Hoover Presidency: A Reappraisal* (Albany, 1974); Murray N. Rothbard, "The Hoover Myth" in *For a New America: Essays in History and Politics from Studies on the Left, 1959-1967*, James Weinstein and David W. Eakins, eds. (New York, 1970), pp. 162-79.

6. For a discussion of general New Deal ideological inconsistency, see Ellis Hawley, *The New Deal and the Problem of Monopoly* (Princeton, 1966).

7. William Appleton Williams, *The Contours of American History* (Chicago, 1966), pp. 428, 439-50.

8. Samuel B. Bledsoe, Interview, Columbia Oral History Library, Special Collections Department, Columbia University, p. 401.

9. *Saturday Evening Post* (January 19, 1946), p. 216.

BIBLIOGRAPHY

Manuscript Collections and Private Papers

Branson, E. C., Southern Historical Collection, University of North
 Carolina, Chapel Hill, N.C.
Carmody, John, Franklin D. Roosevelt Library, Hyde Park, N.Y.
Cooke, Morris Llewellyn, Franklin D. Roosevelt Library, Hyde Park, N.Y.
Ehringhaus, J. C. B., Department of Archives and History, Raleigh, North
 Carolina.
Files of the Secretary of Agriculture, Record Group 16, National Archives.
Hoover, Herbert, Herbert Hoover Library, West Branch, Iowa.
Johnston, Mercer Green, Manuscript Division, Library of Congress.
King, Judson, Manuscript Division, Library of Congress.
Meares, R. A., South Carolinian Library, University of South Carolina,
 Columbia, S.C.
Norris, George W., Manuscript Division, Library of Congress.
Pinchot, Gifford, Manuscript Division, Library of Congress.
Poe, Clarence, Department of Archives and History, Raleigh, North
 Carolina.
Rayburn, Samuel T., Sam Rayburn Library, Bonham, Tx.
Roosevelt, Franklin D., Franklin D. Roosevelt Library, Hyde Park, N.Y.
Rural Electrification Administration, Record Group 221, National Archives.
Slattery, Harry, Manuscript Collection, Duke University, Durham, N.C.

Oral History

Columbia Oral History Collections, Columbia University.
  Bledsoe, Samuel B.
  Carmody, John
  O'Neal, Edward
Sam Rayburn Library, Bonham, Tx.
  McCormack, John W.

Interviews

Bailey, Truett, August 23, 1973, Cleburne, Texas.
Easter, E. C., June 17, 1969, Birmingham, Alabama.
Ellis, Clyde T., August 5-6, 1969, Washington, D.C.
Richey, James B., December 23, 1968, Bonham, Texas.
Wright, Douglas G., June 11, 1969, February 5, 1972, Tulsa, Oklahoma.

Public Documents

Beall, Robert T. "Rural Electrification." *1940 Yearbook of Agriculture.*
  Washington, U.S. Government Printing Office, 1940.
Bureau of the Census, *Census of the Population, 1950.* I. Washington,
  U.S. Government Printing Office, 1952.
———, *Fifteenth Census of the United States, Agriculture,* V. Washington,
  U.S. Government Printing Office, 1932.
———, *Fourteenth Census of the United States, Agriculture,* V. Washing-
  ton, U.S. Government Printing Office, 1922.
———, *Statistical Abstract of the United States, 1935.* Washington, U.S.
  Government Printing Office, 1935.
Congressional Record, 1920-1960.
Cornell, F. D., "Power on West Virginia Farms," *West Virginia Agricultural
  Experiment Station Bulletin* no. 267 (June 1935).
Daniels, A. M., "Electric Light and Power in the Farm House," *1919 Year-
  book of Agriculture,* Washington, D.C., U.S. Government Printing
  Office, 1920.
Electric Home and Farm Authority, *TVA, Electricity for All.* Pamphlet.
  Government Publications Room, University of California, Los Angeles,
  undated.

Georgia Agricultural Experiment Station, USDA Bureau of Agricultural Economics, "Electricity on Farms in the Upper Piedmont of Georgia," *Georgia Agricultural Experiment Station Bulletin* no. 263 (June 1950).

Giant Power, Proceedings before the Committee on Corporations of the Senate and the Manufacturers Committee of the House of Representatives, being a Joint Hearing on Senate Giant Power Bills, Numbers 32-37 (Pennsylvania Assembly, Extraordinary Session of 1926).

Giant Power, Report of the Giant Power Survey Board to the General Assembly. Harrisburg, Pennsylvania, Joint Committee, February 1925.

Meares, R. A., Executive Secretary, Legislative Joint Committee on Rural Electrification, "Report of Investigation Concerning the Progress of Rural Electrification and Development of South Carolina," *Report of State Officers, Boards, and Committees to the General Assembly of South Carolina*, I. Columbia, S.C.: Joint Committee on Printing, 1939.

Mississippi Agricultural Experiment Station, USDA Bureau of Agricultural Economics, "Electricity on Farms in the Clay Hills Area of Mississippi," *Mississippi Agricultural Experiment Station Bulletin* no. 493 (August 1952).

Mississippi Valley Committee, *Report of the Mississippi Valley Committee of the Public Works Administration.* Washington, U.S. Government Printing Office, 1936.

Morgan, Harcourt A., *Rural Electrification, A Promise to American Life.* Pamphlet. Tennessee Valley Authority, Washington, U.S. Government Printing Office, 1936.

Morse, George, "Rural Electrification," *Giant Power Survey Report* (Harrisburg, 1925), pp. 117-40.

North Carolina Committee on Rural Electrification, *North Carolina Rural Electrification Survey.* Raleigh, North Carolina Emergency Relief Administration, 1934.

North Carolina General Assembly, *Public Laws and Resolutions, Session of 1917.* Raleigh, State Printers and Binders, 1917.

Oklahoma Agricultural Experiment Station, "Rural Electrification in Oklahoma," *Oklahoma Experiment Station Bulletin* no. 207 (December 1932).

Rural Electrification Administration, *Annual Reports*, 1935-1956.

——, "Rural Lines USA," *U.S. Department of Agriculture Miscellaneous Publication* no. 811 (March 1966).

———, *Statistical Report, 1953.* Washington, U.S. Government Printing Office, 1953.

———, *Statistical Report, 1954.* Washington, U.S. Government Printing Office, 1954.

Schuler, Edgar A., Louisiana State University Extension Service, "Survey of Radio Listeners in Louisiana," *Louisiana State University Extension Service.* Pamphlet (1943).

South Carolina, 80th General Assembly, 1st Session, *House Journal.* Columbia, S.C., Joint Committee on Printing, February 10, 1932.

———, *House Journal.* Columbia, S.C., Joint Committee on Printing, February 16, 1933.

South Carolina Power Rate Investigation Committee, *Report on the Electric Utility Situation in South Carolina.* Columbia, S.C. Joint Committee on Printing, December 1931.

South Carolina State Highway Commission, *Application of the South Carolina Highway Department for a Loan of $5,912,800 from the Federal Emergency Administration of Public Works.* Columbia, S.C., Joint Committee on Printing, December 1933.

Stewart, E. A., "Rural Electrification in Europe," *University of Minnesota Department of Agriculture Paper* no. 661 (March 1927).

Tennessee Agricultural Experiment Station, USDA Bureau of Agricultural Economics, "Electricity on Farms and Rural Homes in the East Tennessee Valley," *Tennessee Agricultural Experiment Station Bulletin* no. 221 (April 1951).

United States Department of the Interior, *Annual Report, 1943.* Washington, U.S. Government Printing Office, 1943.

———, *Annual Report, 1944.* Washington, U.S. Government Printing Office, 1944.

———, *Annual Report, 1950,* Washington, U.S. Government Printing Office, 1950.

———, Southwestern Power Administration, *Facts about the Southwestern Power Administration.* Tulsa, Okla., Government Printing Office, 1969.

———, Southwestern Power Administration, *A History of the SPA.* Tulsa, Okla., Government Printing Office, 1952.

U.S. Congress, Committee on Appropriations, *Interior Department Appropriations for 1954,* 83d Congress, 1st Sess. Washington, U.S. Government Printing Office, 1953.

U.S. Congress, Committee on Flood Control, *Flood Control on the Mississippi and its Tributaries,* 71st Congress, 2d Sess. Washington, U.S. Government Printing Office, 1930.

U.S. Congress, House, Committee on Interstate and Foreign Commerce, *A Bill to Provide for Rural Electrification and for other Purposes,* 74th Congress, 2d Sess. Washington, U.S. Government Printing Office, 1936.

U.S. Congress, House, Conference Report no. 2219, *Rural Electrification,* 74th Congress, 2d Sess. Washington, U.S. Government Printing Office, 1936.

U.S. Congress, House, Conference Report no. 2644, *Rural Electrification Act of 1936,* 74th Congress, 2d Sess. Washington, U.S. Government Printing Office, 1936.

U.S. Congress, House, Document no. 541, 75th Congress, 3d Sess. Washington, U.S. Government Printing Office, 1938.

U.S. Congress, House, Subcommittee of the Agricultural Committee, *Rural Electrification,* 78th Congress, 1st Sess. Washington, U.S. Government Printing Office, 1944.

U.S. Congress, House, Subcommittee of the Committee on Appropriations, *Interior Department Appropriation Bill for 1947, Southwestern Power Administration,* 79th Congress, 2d Sess. Washington, U.S. Government Printing Office, 1946.

————, *Agricultural Appropriations for 1948,* 80th Congress, 1st Sess. Washington, U.S. Government Printing Office, 1947.

————, *Interior Department Appropriations for 1952,* 82d Congress, 1st Sess. Washington, U.S. Government Printing Office, 1951.

U.S. Congress, House, Subcommittee on Interstate and Foreign Commerce, *Rural Electrification Planning,* 79th Congress, 1st Sess. Washington, U.S. Government Printing Office, 1945.

U.S. Federal Emergency Administration of Public Works, *Non-Federal PWA Power Projects, Allotments Authorized Under the National Industrial Recovery Act of 1933, Emergency Appropriation Act, Fiscal Year, 1935, Emergency Relief Appropriation Act of 1935, and the First Deficiency Appropriation Act, Fiscal Year 1936,* Washington, U.S. Government Printing Office, 1937.

U.S. Congress, Senate, Subcommittee of the Committee on Agriculture and Forestry, *Administration of the Rural Electrification Act,* 78th Congress, 2d Sess. Washington, U.S. Government Printing Office, 1944.

U.S. Congress, Senate, *Agricultural Appropriations for 1948*, 80th Congress, 1st Sess. Washington, U.S. Government Printing Office, 1947.

———, Committee on Agriculture and Forestry. *Hearings on S. Res. 197, a resolution authorizing the employment of assistants and the expenditure of funds in a proposed investigation of the administration of the Rural Electrification Act*, 78th Congress, 1st Sess. Washington, U.S. Government Printing Office, 1943-44.

———, *Planning of Rural Electrification Projects*, 79th Congress, 1st Sess. Washington, U.S. Government Printing Office, 1945.

Waggoner, J. E., "Electricity on Texas Farms," *Texas Agricultural and Mechanical College Bulletin* no. 35 (February 1, 1928).

Wellwood, A. R., State Highway Commission, *Survey of Rural Electrification in South Carolina*. Columbia, S.C., Joint Committee on Printing, 1933.

Important Documents in Manuscripts and Other Collections

Alabama Farm Bureau Federation, *Progress on Rural Electric Service in Alabama*. Pamphlet (1925), files of Alabama Power Company.

Alabama Power Company, *Powergrams*, 6 (October 1926), files of Alabama Power Company.

Brunner, Edward S., *Radio and the Farmer*. Pamphlet. Radio Institute of the Audible Arts (New York, 1936), National Agricultural Library.

Carroll, Mollie Ray, "General Conclusions From a Survey of the Moral and Cultural Effects of the Inside Bathroom and the Refrigerator" (1935). Conducted in Two Counties Each of Virginia, Ohio, and Minnesota." Typescript.

"Conference of Cooperative Representatives and Rural Electrification Administration" (Washington, June 6, 1935). Typescript, National Agricultural Library.

Cooke, Morris L., "National Plan for the Advancement of Rural Electrification Under Federal Leadership and Control with State and Local Cooperation and as a Wholly Public Enterprise" (1934). Typescript, Library of Congress.

———, "Report of the Rural Electrification Administration to the National Emergency Council" (October 14, 1935). Typescript, National Agricultural Library.

Easter, E. C., M. J. Funchess, and M. L. Nichols, Department of Agricultural Engineering, Alabama Polytechnic Institute, "Progress Report of Project in Rural Electricity" (1925). Typescript, files of Alabama Power Company.

Electric Home and Farm Authority, "Purposes and Program with Recommendations for Expansion," Interdepartmental Report (March 15, 1935). Typescript, TVA Library.

Falack, Edward, "Operations of Alcorn County Electric Cooperative" (January 1935). Typescript, National Agricultural Library.

Georgia Power Company, "Rural Electric Service in Georgia" (Atlanta, June 27, 1936). Typescript, Record Group 221, National Archives.

Hobbs, S. H., "A Brief History of Rural Electrification in North Carolina." Unpublished Presidential Address, North Carolina Historical Society (November 1, 1963). Typescript, North Carolina Room, University of North Carolina.

Jackson, C. W., "A Study of the Value of the Texas Farm and Home Radio Program in Twenty Northeast Texas Counties," (May 1946). Typescript, National Agricultural Library.

Jordon, W. Fred, *The Arkansas Plan*, REA Interdepartmental Pamphlet (June 1940), National Agricultural Library.

Kefauver, Nora M., "Rural Electric Lighting in Areas of Newly Energized Lines, Lafollette Electric Department, Lafollette, Tennessee, Extending into Campbell and Claiborne Counties" (March 30, 1942). Division of Electrical Development, Tennessee Valley Authority, Mimeographed, TVA Library.

King, Judson, *National Popular Government League Bulletin* no. 206 (July 23, 1943), National Agricultural Library.

———, "Special Bulletin to Board Members of REA Cooperatives," *National Popular Government League Bulletin* (July 4, 1943), National Agricultural Library.

———, "Who Will Get the $100,000,000 for Farm Electrification?" *National Popular Government League Bulletin* no. 171 (April 25, 1935). National Agricultural Library.

Mamer, Louisan, REA Home Economics Department, "Electricity Pays Its Own Way in the Rural Home" (March 11, 1952). Typescript, National Agricultural Library.

Meares, R. A., "Proposed Investigation of Public Utilities in the State of South Carolina" (1929). Typescript, Meares Papers.

Munger, G. D., Electric Home and Farm Authority, "Methods Employed for Financing Equipment and Appliance Purchases." Paper read at the Third World Conference (September 1936). Typescript, TVA Library.

National Resources Board, *A Report on National Planning and Public Works in Relation to Natural Resources and Including Land Use and Water Resources with Findings and Recommendations.* Washington, U.S. Government Printing Office, 1934.

Norcross, T. W., "A New Deal in Rural Electrification" (May 1935). Typescript, National Agricultural Library.

North Carolina Bureau of Extension, University of North Carolina, *News Letter,* 1914-1936, North Carolina Room, University of North Carolina Library.

Prance, Bill, "Agricultural Activities of Radio Station WSB, June 17, 1940 through December 31, 1940," (Atlanta, 1941). Typescript, National Agricultural Library.

Rommel, George M., "Rural Electrification in the South" (1934). Typescript, TVA Library.

Rural Electrification Administration, Interbureau Committee on Rural Electrification, "Effects of Use of Electricity on Agriculture and Farm Life" (October 1941). Typescript, Johnston Papers.

———, Postwar Planning Committee, *Rural Electrification After the War,* A Preliminary Report (St. Louis, 1944). Pamphlet, National Agricultural Library.

———, "Present Uses of Electricity in Rural Areas" (1941). Typescript, Johnston Papers.

———, *Rural Electrification News,* 1935-1954.

"Rural Electrification in the South," *Progressive Farmer.* Special Report (1945). Mimeographed, files of Alabama Power Company.

Saville, Allan J., "Preliminary Survey of Rural Electrification in Virginia for State Corporation Commission" (March 1, 1935). Typescript, National Agricultural Library.

Stewart, E. A., *Electricity in Rural Districts Served by the Hydro-Electric Power Commission of the Province of Ontario, Canada.* (1926). Pamphlet, files of Alabama Power Company.

Texas Power and Light Company, *Electricity and Agriculture.* (October 1927). Pamphlet, files of Alabama Power Company.

The Browning TVA Rural Electrification Plan (Nashville, 1938). Pamphlet, TVA Library.

"Typescript of the First Meeting of the NRECA in Washington, D.C., 1944." Typescript, Record Group 221, National Archives.
White, E. A., "What's Going on in the South" (undated). Typescript, files of Alabama Power Company.

Publications of the National Electric Light Association (NELA) and the Committee on the Relation of Electricity to Agriculture (CREA)

National Electric Light Association, *Annual Proceedings,* 1920-1935.
———, *Bulletin,* 1924-1933.

Pamphlets of the NELA

Developing Electric Service for the Farms as Exemplified by the Alabama Power Company, Publication no. 289-65 (July 1929), files of Alabama Power Company.
Progress in Rural and Farm Electrification, 1921-1931, Publication no. 237 (August 1932), files of Alabama Power Company.
Report of the Committee on Electricity in Rural Districts (Chicago, 1911), National Agricultural Library.

Pamphlets of the CREA

Alabama Project, The (November 1924), files of Southern California Edison Electric Company.
Electricity on the Farm and in Rural Communities (November 1931), files of Alabama Power Company.
Rural Electric Service in Alabama (April 1926), files of Southern California Edison Electric Company.
Rural Electrification in Virginia (May 1926), files of Alabama Power Company.
Sixth Annual Report (Chicago 1929), files of Alabama Power Company.
Tenth Annual Report (Chicago 1933), files of Alabama Power Company.

Papers, Speeches

Corbett, R. B., "What National Farmers' Organizations Should be Doing about Farm Electrification." Speech to National Farm Electrification Conference, Chicago, November 7-8, 1946. Typescript, National Agricultural Library.

Kable, George, "Rural Electrification in Alcorn County, Mississippi." Paper presented at the American Society of Agricultural Engineers, Athens, Georgia, June 20, 1935, TVA Library.

Lilienthal, David, "Future of Farm Electricity" (November 18, 1939), TVA Library.

———, "Progress in the Electrification of the American Home and Farm" (September 19, 1934), TVA Library.

———, "The Tennessee Valley Authority and Farm Electrification" (December 12, 1934), TVA Library.

McAlister, J. T., "Agricultural Engineering, Its Origin and Growth in South Carolina." Paper presented at the First Annual Meeting of the South Carolina Section of the American Society of Agricultural Engineers, Clemson, South Carolina, November 22, 1957, Clemson University Library.

Martin, Thomas W., "Remarks on Beginning of Rural Service to Pittsville, Alabama (August 21, 1935), files of Alabama Power Company.

Weaver, David S., "A Rural Electrification Survey." Paper presented at the American Society of Agricultural Engineers, Athens, Georgia, June 17-20, 1935, TVA Library.

Newspapers

Atlanta Constitution, 1935.

Barron's National Financial Weekly, 1935.

Bonham (Tex.) *Daily Favorite,* 1923-1936, 1957, 1960.

Charleston News and Courier, 1933-1934.

Chattanooga News, 1935-1936.

New York Times, 1918-1955.

Raleigh News and Observer, 1932-1936.

Sherman (Tex.) *Daily Democrat,* 1938, 1943.

Clippings—1933-1955—from newspaper file of TVA Library.

Dissertations, Theses

Christie, Jean, "Morris Llewellyn Cooke: Progressive Engineer." Doctoral dissertation, Columbia University, 1963.

Cutler, Addison T., "Rural Electrification: A Sketch of the Problems Involved in Bringing Electricity to the Farm, with Special Reference to North Carolina." Masters thesis, University of North Carolina, 1926.

Deutsch, Joseph, "Rural Electrification in North Carolina." Masters thesis, University of North Carolina, 1944.

Green, Michael Knight, "A History of the Public Rural Electrification Movement in Washington to 1942." Doctoral dissertation, University of Idaho, 1967.

Hobson, Leo G., "The Agricultural Cooperatives and Rural Electrification." Doctoral dissertation, Cornell University, 1936.

Sanders, Albert N., "State Regulation of Public Utilities by South Carolina, 1879-1935." Doctoral dissertation, University of North Carolina, 1956.

Stauter, Mark C., "The Rural Electrification Administration, 1935-1945, A New Deal Case Study." Doctoral dissertation, Duke University, 1973.

Tyler, Delbert Earl, "History of the Fannin County Rural Electric Cooperative." Masters thesis, East Texas State University, 1964.

Books

Albertson, Dean, *Roosevelt's Farmer: Claude R. Wickard in the New Deal.* New York, Columbia University Press, 1955.

Anderson, Oscar E., *Refrigeration in America: A History of a New Technology and its Impact.* Princeton, Princeton University Press, 1953.

Barnes, Joseph, *Wilkie.* New York, Simon and Schuster, 1952.

Bauer, John, *The Electrical Power Industry: Development, Organization and Public Bodies.* New York, Harper and Brothers, 1939.

Campbell, Christina, *The Farm Bureau and the New Deal: A Study of the Making of National Farm Policy, 1933-1940.* Urbana, University of Illinois Press, 1962.

Childs, Marquis, *The Farmer Takes a Hand,* Garden City, N.Y., Doubleday, 1953.

————, *Sweden, The Middle Way.* New Haven, Yale University Press, 1936.

Clapp, Gordon, *The TVA, An Approach to the Development of a Region.* Chicago, University of Chicago Press, 1955.

Clark, Thomas, *The Emerging South.* New York, Oxford University Press, 1961.

Cooke, Morris, ed., *What Electricity Costs.* New York, New Republic, 1933.

Corbitt, David L., ed., *Addresses, Letters and Papers of John Christopher Blucher Ehringhaus, Governor of North Carolina, 1933-1937.* Raleigh, Council of State, 1950.

Crawford, Geddings Hardy, ed., *Who's Who in South Carolina.* Columbia, S.C., McGraw, 1921.

Dabney, Virginius, *Below the Potomac.* New York, D. Appleton-Century, 1942.

Daniels, Jonathan, *White House Witness, 1942-1945.* Garden City, N.Y., Doubleday, 1975.

Dorough, Dwight, *Mr. Sam.* New York, Random House, 1962.

Eastman, E. R., *These Changing Times.* New York, Macmillan, 1927.

Ehlers, Victor and Ernest Steel, *Municipal and Rural Sanitation.* New York, McGraw-Hill, 1943.

Ellis, Clyde T., *A Giant Step.* New York, Random House, 1966.

Evans, James E., *Prairie Farmer and WLS: The Burridge D. Butler Years.* Urbana, University of Illinois Press, 1969.

Fausold, Martin and George T. Mazuzan, eds., *The Hoover Presidency: A Reappraisal.* Albany, State University of New York Press, 1974.

Fletcher, Stevenson W., *Pennsylvania Agriculture and Country Life, 1846-1940.* Harrisburg, Pennsylvania Historical and Museum Commission, 1955.

Funigiello, Philip, *Toward a National Power Policy: The New Deal and the Electrical Industry.* Pittsburgh, University of Pittsburgh Press, 1973.

Garwood, John D., *The Rural Electrification Administration, An Evaluation.* Washington, American Enterprise Institute, 1963.

Grubbs, Donald H., *Cry From Cotton: The Southern Tenant Farmers' Union and the New Deal.* Chapel Hill, University of North Carolina Press, 1971.

Hawley, Ellis W., *The New Deal and the Problem of Monopoly.* Princeton, Princeton University Press, 1966.

Hubbard, Preston, *Origins of the TVA.* Nashville, Vanderbilt University Press, 1961.

Ickes, Harold, *Back to Work, the Story of PWA.* New York, Macmillan, 1935.

Kile, O. M., *The Farm Bureau Through Three Decades.* Baltimore, Waverly Press, 1948.

King, Judson, *The Conservation Fight: From Theodore Roosevelt to the Tennessee Valley Authority.* Washington, Public Affairs Press, 1959.

Kirkendall, Richard, *Social Scientists and Farm Politics in the Age of Roosevelt.* Columbia, University of Missouri Press, 1966.

Knapp, Joseph G., *The Advance of the American Cooperative, 1920-45.* Danville, Ill., Interstate Printers and Publication, Inc., 1973.

Kolko, Gabriel, *The Triumph of Conservatism: A Reinterpretation of American History, 1900-1916.* Chicago, Quadrangle, 1967.

Leuchtenburg, William, *Franklin D. Roosevelt and the New Deal.* New York, Harper & Row, 1963.

Lilienthal, David E., *The Journals of David E. Lilienthal, Vol. I: TVA Years, 1932-1945.* New York, Harper & Row, 1964.

——, *TVA, Democracy on the March.* New York, Harper & Brothers, 1944.

Lincoln, Murray, *Vice-President in Charge of Revolution.* New York, McGraw-Hill, 1960.

Lowitt, Richard, *George Norris, The Making of a Progressive, 1861-1912.* Syracuse, Syracuse University Press, 1963.

——, *George Norris: The Persistence of a Progressive, 1913-1945.* Urbana, University of Illinois Press, 1971.

McGraw, Thomas K., *TVA and the Power Fight 1933-1939.* Philadelphia, J. B. Lippincott, 1971.

McGreary, M. Nelson, *Gifford Pinchot, Forester, Politician.* Princeton, Princeton University Press, 1960.

Martin, Thomas W., *The Story of Electricity in Alabama.* Birmingham, Alabama Power Company, 1953.

Moody's Public Utility Manual. New York, Moody's Investor Service, 1955.

Mott, Frederick D. and Milton I. Roemer, *Rural Health and Medical Care.* New York, McGraw-Hill, 1948.

Muller, Frederick, *Public Rural Electrification.* Washington, American Council Public Affairs, 1944.

Murray, William S., *Government Owned and Controlled Compared with Privately Owned and Regulated Electric Utilities in Canada and the United States.* New York, National Electric Light Association, 1922.

National Rural Electric Cooperative Association, *Rural Electric Fact Book.* Washington, NRECA, 1965.

Noggle, Burl, *Teapot Dome: Oil and Politics in the 1920's.* Baton Rouge, Louisiana State University Press, 1962.

Norris, George, *Fighting Liberal.* New York, Macmillan, 1945.

Odum, Howard, *Southern Regions.* Chapel Hill, University of North Carolina Press, 1936.

Pinckett, Harold T., *Gifford Pinchot: Public and Private Forester.* Urbana, University of Illinois Press, 1970.

Rankin, John, "Rural Electrification by Power County Associations," *The Reference Shelf, Government Ownership of Electric Utilities,* ed. Julia E. Johnson. New York, Wilson Company, 1936, pp. 229-35.

Richardson, L. K., *Wisconsin REA: The Struggle to Extend Electricity to Rural Wisconsin 1935-1955.* Madison, University of Wisconsin Experiment Station, 1961.

Schlesinger, Arthur, M., Jr., *The Politics of Upheaval.* Boston, Houghton-Mifflin, 1960.

Slattery, Harry, *Rural America Lights Up.* Washington, National Home Library Foundation, 1940.

Steinberg, Alfred, *Sam Johnson's Boy, A Close-Up of the President from Texas.* New York, Macmillan, 1968.

Trombley, Kenneth, *The Life and Times of a Happy Liberal, Morris Llewellyn Cooke.* New York, Harper & Brothers, 1954.

Twentieth Century Fund, *Electric Power and Government Policy: A Survey of the Relations between Government and the Electric Power Industry.* New York, Lord Baltimore Press, 1948.

Voorhis, Jerry, *American Cooperatives.* New York, Harper & Row, 1961.

Weinstein James and David W. Eakins, eds., *For a New America: Essays in History and Politics from Studies on the Left, 1959-1967.* New York, Random House, 1970.

Wiebe, Robert E., *Businessmen and Reform: A Study of the Progressive Movement.* Chicago, Quadrangle, 1968.

Williams, William Appleton, *The Contours of American History.* Chicago, Quadrangle, 1966.

Wilson, Louis R., *The University of North Carolina, 1900-1930.* Chapel Hill, University of North Carolina Press, 1957.

Articles

"Accomplishments in Rural Electric Development by Alabama Power Company," *Edison Electric Institute Bulletin,* 2 (January 1934): 9-17.

American Federationist, 32 (April 1935): 233-34.

Ball, Thomas F., "State to Build and Operate Its Own Rural Electric Lines," *Public Utilities Fortnightly,* 12 (June 1934): 772-80.

Baker, Gladys, "And to Act for the Secretary: Paul H. Appleby and the Department of Agriculture, 1933-1940," *Agricultural History,* 45 (October 1971): 235-38.

Bates, J. Leonard, "Fulfilling American Democracy: The Conservation Movement, 1907-1921," *Mississippi Valley Historical Review,* 44 (June 1957): 29-57.

Beeler, M. N., "Why Alabama Leads in Electrifying Farms," *Cappers' Farmer,* 38 (April 1927): 7-8.

Bradford, Ernest, "The Influence of Cheap Power on Factory Location and on Farming," *Annals of the American Academy of Political and Social Science,* 117 (March 1925): 91-95.

Brown, Deward C., "The Sam Rayburn Papers: A Preliminary Investigation," *The American Archivist,* 35 (July-October 1972): 331-36.

Christie, Jean, "Giant Power: A Progressive Proposal of the Nineteen-Twenties," *Pennsylvania Magazine of History and Biography,* 96 (October 1972): 480-507.

———, "The Mississippi Valley Committee: Conservation and Planning in the New Deal," *The Historian,* 32 (May 1970): 449-69.

Collins, James H., "Sober Sense about Super-Power," *The Nation's Business,* 12 (March 1924): 21-23.

Cooke, Morris, "The Early Days of the Rural Electrification Idea: 1914-1936," *American Political Science Review,* 42 (June 1948): 431-47.

———, "The Long Look Ahead," *The Survey,* 51 (March 1, 1924): 21-23.

———, "A Note on Rates for Rural Electric Service," *Annals of the American Academy of Political and Social Science,* 118 (March 1925): 52-59.

Current Opinion, 83 (August 1922): 236-37.

Easter, E. C., "History of Rural Electrification in Alabama," *Alabama Agricultural Engineer,* 1 Supplement (January 1954): 4-6.

Electrical World, 1920-1955.

Erdman, H. E., "Some Social and Economic Aspects of Rural Electrification," *Journal of Farm Economics,* 12 (April 1930): 311-19.

Evans, Harold, "The World's Experience with Rural Electrification," *Annals of the American Academy of Political and Social Science,* 118 (March 1925): 30-42.

Farrar, Lorstan D., "The Rural Electric Co-ops Get a Trade Association," *Public Utilities Fortnightly,* 31 (May 27, 1943): 676-82.

Funchess, M. J., "The Agricultural College and Rural Electrification," *Powergrams, Alabama Power Company,* 6 (October 1926): 3-5.

Funigiello, Philip J., "Kilowatts for Defense: The New Deal and the Coming of the Second World War," *Journal of American History,* 56 (December 1969): 604-20.

Gompers, Samuel, "Giant Power—Its Possibilities, Potentialities and its Administration," *American Federationist,* 30 (December 1923): 570-71.

Gray, John H., "Giant Power: The Giant Power Report of Pennsylvania," *National Municipal Review,* 15 (March 1926): 165-172.

Grimes, Katherine, "Light Comes to Alabama," *Southern Agriculturalist,* 59 (September 15, 1929): 5.

Harris, Agnes E., "What Electricity Meant to the Farm," *Powergrams, Alabama Power Company,* 6 (October 1926): 17-18.

Hurd, C. J., "Farm and Community Refrigeration in the South," *Refrigerating Engineering,* 36 (November 1938): 10-14.

——, "Progress of Farm Electrification in the Tennessee Valley," *Agricultural Engineering,* 19 (November 1938): 493-95.

King, Judson, "What are the True Origins of the NRECA," *Public Utilities Fortnightly,* 32 (July 1943): 3-14.

McComb, William, The Steady Demand for Rural Electrification," *Public Utilities Fortnightly,* 16 (September 26, 1935): 428-34.

McCrory, S. H., "Problems Involved and Methods Used in Promoting Rural Electrification," *Journal of Farm Economics,* 12 (April 1930): 320-25.

Marple, Warren H., "An Appraisal of Edison Electric Institute's Statistics on Farm Electrification," *Journal of Land and Public Utility Economics,* 14 (November 1938): 471-76.

Moore, Harry E. and Bernice M. Moore, "Problems of Reintegration of Agrarian Life," *Social Forces,* 15 (March 1937): 384-90.

National Municipal Review, 15 (March 1926): 165-72.

New Republic (February 16, 1947), p. 34.

New Republic (June 21, 1922), pp. 102-04.

Nichols, M. L., "Research Work and Rural Electrification," *Powergrams, Alabama Power Company,* 6 (October 1926): 11-12.

Pattison, Mary, "The Abolition of Household Slavery," *Annals of the American Academy of Political and Social Science,* 118 (March 1925): 124-27.

Person, H. S., "The Rural Electrification Administration in Perspective," *Agricultural History,* 24 (April 1950): 70-89.

Progressive Farmer, 1918-1935.

Rau, O. M., "Rural Electric Lines to Curb Migration from Farms,"
   Public Utilities Fortnightly, 16 (August 15, 1935): 200-09.
"Rural Electrification, A Postwar Market Forecast," *Country Gentleman*,
   114 (Philadelphia, 1944).
Rural Electrification Exchange (April 1938): 33-35.
Rural Electrification Exchange (October 1941): 73-74.
Rural Electrification Exchange (January 1947): 10.
Rural Electrification Exchange (November 1955): 24-25.
Saville, Allan J., "The Virginia Plan," *Public Utilities Fortnightly*, 16
   (September 12, 1935): 312-16.
Scientific American Monthly, 1 (May 1920): 473.
Shideler, James H., "Flappers and Philosophers and Farmers: Rural-
   Urban Tensions of the Twenties," *Agricultural History*, 47 (October
   1973): 283-99.
Stewart, Royden, "Bringing Power to the Farm: Early Development
   Years," *Public Utilities Fortnightly*, 27 (May 8, 1941): 579-87.
———, "Bringing Power to the Farm: National Development 1924-1935,
   The CREA," *Public Utilities Fortnightly*, 27 (May 22, 1941): 651-63.
———, "Rural Electrification in the United States: The Pioneer Period,
   1906-1923," *Edison Electric Institute Bulletin*, 9 (September 1941):
   382-83.
U.S. News and World Report, 32 (June 6, 1952): 56.
Weaver, David S., "Story of Rural Electrification in North Carolina,"
   Public Utilities Fortnightly, 19 (June 1937): 808-15.
White, E. A., "The Steady Demand for Rural Electrification," *Public
   Utilities Fortnightly*, 16 (August 29, 1935): 265-69.
Wilkes, Frank M., "It CAN Happen Here—and it HAS!" *Electrical South*,
   33 (July 1953): 18-21.

INDEX

ABOUT THE AUTHOR

D. Clayton Brown is Associate Professor of History at Texas Christian University in Fort Worth, Texas. He is the author of *Rivers, Rockets, and Readiness: Army Engineers in the Sunbelt,* in addition to several articles in scholarly journals.